UNIVERSITY OF CAMBRIDGE
DEPARTMENT OF APPLIED ECONOMICS

MONOGRAPH 27

WAGES POLICY IN THE BRITISH COALMINING INDUSTRY

A STUDY OF NATIONAL WAGE BARGAINING

UNIVERSITY OF CAMBRIDGE
DEPARTMENT OF APPLIED ECONOMICS

Monographs

This series consists of investigations conducted by members of the Department's staff and others working in direct collaboration with the Department.

The Department of Applied Economics assumes no responsibility for the view expressed in the Monographs published under its auspices.

5. The Lognormal Distribution
by J. AITCHISON *and* J. A. C. BROWN
8. British Economic Growth, 1688–1959
by PHYLLIS DEANE *and* W. A. COLE (*Second edition*)
13. The British Patent System 1. Administration *by* KLAUS BOEHM
15. Planning, Programming and Input–Output Models
Selected Papers on Indian Planning *by* A. GHOSH
16. Biproportional Matrices and Input–Output Change
by MICHAEL BACHARACH
17. Abstract of British Historical Statistics
by B. R. MITCHELL *in collaboration with* PHYLLIS DEANE
18. Second Abstract of British Historical Statistics
by B. R. MITCHELL *and* H. G. JONES
21. British Monetary Policy 1924–31
by D. E. MOGGERIDGE
22. Programming and Inter-regional Input–Output Analysis
by A. GHOSH
23. The Economic Impact of the Patent System
by C. T. TAYLOR *and* Z. A. SILBERSTON
24. Planning Problems in the USSR
by M. ELLMAN
25. Technological Diffusion and the Computer Revolution: The UK Experience
by PAUL STONEMAN
26. Industrial Pricing in the United Kingdom
by KENNETH COUTTS, WYNNE GODLEY
and WILLIAM NORDHAUS

WAGES POLICY IN THE BRITISH COALMINING INDUSTRY

A STUDY OF NATIONAL WAGE BARGAINING

L. J. HANDY

Head of the Industrial Relations Faculty
Ashridge Management College

CAMBRIDGE UNIVERSITY PRESS

CAMBRIDGE

LONDON NEW YORK NEW ROCHELLE

MELBOURNE SYDNEY

Published by the Press Syndicate of the University of Cambridge
The Pitt Building, Trumpington Street, Cambridge CB2 1RP
32 East 57th Street, New York, NY 10022, USA
296 Beaconsfield Parade, Middle Park, Melbourne 3206, Australia

First published 1981

Photoset and printed in Malta by Interprint Limited

British Library Cataloguing in Publication Data
Handy, L. J.
Wages policy in the British coalmining industry.
– (University of Cambridge. *Department of Applied Economics*. Monographs; 27).
1. Wages – Coal miners – Great Britain
2. National Coal Board
331.2′15 HD4966.M63G7 80-40229

ISBN 0 521 23535 9

TO MY MOTHER
AND IN MEMORY OF
MY FATHER

CONTENTS

Part Two: National wage decisions and wage structure

TABLES

FIGURES

PREFACE

This book is concerned with the evolution of wages policy in the British coalmining industry since nationalisation, as a case study of a deliberate attempt to revise a wage structure in accordance with central objectives under the pressure of institutional, economic and technological change. The work has been undertaken with two principal objectives in mind. The first is to evaluate central wages policy in coalmining in the light of its professed objectives and the industry's changing circumstances. The second is to examine the relevance of the coalmining industry's wages experience to some broader issues of wages policy.

The book derives principally from the doctoral thesis which I submitted to the University of Cambridge in 1977. There have been some new developments on the coalmining industry's wages front since then, the most important of which is the introduction of productivity-related wage incentive schemes. As this latter development is deserving of special attention and has not caused me to change my original conclusions, I have chosen to update my analysis in a postscript, which deals specifically with the productivity schemes.

I have several acknowledgements to make. The first is to Professor H. A. Turner who, as both a supervisor and colleague, was primarily responsible for developing and guiding my early interest in labour economics and industrial relations. A great deal of the analysis in this book builds on Professor Turner's published work in related areas, as well as his constructive criticism of my own shortcomings. Professor Turner's expertise, however, was generously supplemented by that of Professor Malcolm R. Fisher, now at the University of New South Wales. Although he would be the first to deny it, Professor Fisher provided many subtle insights which have had an important bearing on the shape of this volume as well as the kind of support which must fall outside the normal call of duty. No-one could have been better mentors than these two gentlemen.

The University of Cambridge Department of Applied Economics provided the finance and facilities necessary to undertake the work upon which this study is based, and for this I should like to thank Professor Brian Reddaway and Wynne Godley who, in turn, directed the DAE while I was employed there. I am also grateful to several

DAE colleagues – Ken Coutts, Professor Dudley Jackson (now at Aston University), Barry Moore, Cliff Pratten, Roger Tarling, Frank Wilkinson and Margaret Clark – who assisted me in various ways during the course of the project. John Llewelyn (now at the OECD) also provided invaluable advice during the final stage of the research.

I left Cambridge in February 1977 to join the staff of Ashridge Management College. While the settling-in process that goes with a change in employment has inevitably meant some delay in publication, I have subsequently been able to avail myself of a new range of talents and perspectives. The encouragement I have received at Ashridge has been of immense value to me and in this context I should especially like to thank Philip Sadler, Bernard Barry, Keith Carby and Thelma Quince.

The study relies almost entirely on hitherto unpublished wage data made available by the National Coal Board. In this respect, special mention must be made of the unstinting help given by those people employed in the Industrial Relations, Statistics and Finance Departments at both Hobart House and the Area Offices of the National Coal Board, who, in addition to providing data, answered numerous queries concerning its compilation. I should also like to thank the National Union of Mineworkers and its constituent associations for making union records available upon request and for affording me hospitality at their Annual Conferences.

Statistics and records, of course, only tell part of any story and for this reason I spent a large proportion of my time in the coalfields talking with the men, managers, and union officials who are responsible for the day-to-day affairs of the industry. While I cannot mention everyone by name, the wealth of information gleaned from such conversations has proved invaluable. Many of the views expressed in this study have received a good airing at miners' week-end schools up and down the country, and for this opportunity I am grateful to the various unions and Workers' Education Authority branches who extended the invitations to me.

In addition to the help I have received from the people and organisations mentioned above, several other people have supplied useful comments on the text. Derek Robinson, of the Oxford University Institute of Statistics, found time in a busy work schedule as Chairman of the Social Sciences Research Council, to comment in detail on Chapters 1–12; as did Bob Rowthorn, of King's College, Cambridge. Lawrence Daly, General Secretary of the National Union of Mineworkers, and Peter Knowles of the Industrial Relations Department of the National Coal Board, also put valuable time aside to read the whole manuscript and draw my attention to several errors. I am also grateful to an anonymous referee, who submitted exceptionally useful comments on matters of statistical presentation and consistency. Ms Jo Bradley read, and made several stylistic improvements to, the penultimate draft. I am especially indebted to Francis Brooke of Cambridge University Press who edited the final typescript and steered the book through the printing stage. Last but by no means least, I should like to thank Mrs D. French, of Fulbourn, Cambridge, who typed the bulk of the script with great patience and understanding when it was needed most, and Helen Davies, who assisted me with the index. Together these people have saved me from committing many errors. Any deficiencies, of either fact or interpretation, that remain, are, of course, my own responsibility.

Ashridge Management College
July 1979

ABBREVIATIONS USED IN THE TEXT

COSA	Colliery Officials Staff Association
D of E	Department of Employment
EMS	Earnings per manshift
FDWB	Five-day-week bonus
GMWU	General and Municipal Workers Union
JNNC	Joint National Negotiating Committee
MFGB	Miners Federation of Great Britain
NACODS	National Association of Colliery Overmen, Deputies and Shotfirers
NBPI	National Board for Prices and Incomes
NCB	National Coal Board
NPLA	National Power Loading Agreement
NUM	National Union of Mineworkers
OECD	Organisation for European Co-operation and Development
OMS	Output per manshift
ROLF	Remotely Operated Longwall Faces
T & GWU	Transport and General Workers Union
WPIS	Weekly paid industrial staff

CURRENCY UNITS

Where reference is made to wage agreements concluded before decimalization old currency units of £ s d (e.g. 90/11 for £4 10s 11d) are used in the text.

When statistical tables on wage levels span pre- and post-decimilization years, earnings are expressed in decimal units.

WEIGHTS AND MEASURES

1 tonne = 0.9842 tons
1 ton = 20 cwts
1 cwt = 112 lbs

INTRODUCTION

The recent past has seen a considerable number of studies of wage-payment systems and their effects in terms of pay levels. And as the number of studies increases so does the impression that those procedures and the wage systems associated with them are in need of a radical overhaul. The wage systems and the pay structure to which they give rise often seem riddled with inconsistencies, unamenable to central directives and prone to encourage industrial unrest as well as wage inflation. The evidence frequently points to outdated methods of payment, bearing little or no relation to the work process, an excessive use of overtime to maintain pay differentials distorted by local manipulations, inverted skill differentials which act as a drag on skill acquisition, and an absence of effective procedures to regulate pay in the place of work, all of which lead to leap-frogging pay claims at both the local and national level.

But, while there seems to be considerable agreement on the need to revise wage systems in many parts of British industry, there is less agreement on how this should be achieved and the direction in which such a reform should move. Some feel that the solution to the problem resides in the workplace itself, and that collective bargaining should become more decentralised to facilitate the necessary changes. Others feel that our pay problems can best be solved within a national framework which permits a greater degree of coordination between pay in the various sectors. As a general proposition, however, this writer would contend that there is no universal approach to these problems and that the solutions depend very much on the circumstances confronting trade unions and employers in the various sectors of British industry.

The writer's own interest in coalmining wages policy stemmed from the deliberate attempt by the principal parties to collective bargaining in the industry – the National Coal Board and the National Union of Mineworkers – to revise the miners' pay structure according to central objectives in what has been a rapidly changing institutional, economic and technological environment. The purpose of that revision was to bring miners' pay within the orbit of central regulation and eliminate anomalies which had previously been a divisive force in the industry. Within that centralised framework, the

NCB and NUM sought to establish greater uniformity of pay than had existed hitherto between men performing similar types of work in the various coalfields and reduce wage disparities between the highest- and lowest-paid skills. Of special importance in this overall strategy was the decision to abolish payment-by-results systems in favour of uniform day wages. The end result of this wage revision has been one of the most highly centralised pay-determining systems in British industry.

Wages policy in coalmining and the wage reforms, in particular, reflect a concerted attempt by the NCB and the NUM to adapt to a changing environment. Most important in this respect were the creation of a single employer negotiating with one national trade union, the need to contain labour costs in a changing market situation, and the rapid technological developments which have occurred in the industry since nationalisation. But the fact remains that these developments have occurred in what is still a very large, geographically dispersed and diverse industry, and one in which there has been a long history of jealously guarded local autonomy. Thus, just as there were many factors conducive to a greater coordination of collective bargaining, there were others which we might have expected to complicate the task of central wage administration. Clearly, then, wages policy in the coalmining industry raises a number of questions which are relevant to the more general wages policy debate. These include the tendency of the wage system towards local distortion and wage drift, the effect of abolishing piecerate-payment systems on productivity and industrial relations, and the sensitivity of labour supply to the drastic changes in wage differentials which accompanied the wage reforms. These, then, are the general questions upon which we hope to shed some light in this study of miners' pay since nationalisation.

After a chapter which outlines some salient features of the environment for wage determination, against which wages policy in coalmining must be understood, this study is divided into four parts. In the first part, which consists of four chapters, we trace the evolution of the industry's policy towards the wage structure. These chapters cover the problems facing the NCB and NUM in wages policy formulation, and the events preceding those national agreements which sought to reform the wage structures of daywagemen and then of pieceworkers to establish a single rate for similarly graded jobs throughout the industry. The last of these principal agreements, which concerned pieceworkers working behind the coalface, was concluded in 1971 – some fifteen years after the first reform which concerned daywagemen. The final chapter in this part brings the reader up to date on national developments in the daywage structure after 1955. As well as outlining the main features of central wage reforms, this section serves to illustrate that, despite having moved in a generally egalitarian direction, wages policy, in fact, evolved in a very pragmatic fashion which revealed a rather unhappy compromise between equity on the one hand and economic and political expediency on the other.

The second part is concerned with the impact of national wages policy – as outlined in the first part – on wage levels and movements. The first of the three chapters which comprise this evaluation examines in some detail the impact of central wage decisions on general wage movements. This analysis proceeds with an account of the way in which national agreements could have affected miners' pay since nationalisation and, through a quantitative analysis of wage composition, estimates their contribution to the aggregate wage movement as well as that of major groups of mineworkers.

Our approach enables us to measure directly the contribution of such elements as overtime and the nationally agreed pay supplements to wage movements as well as the effect of such factors as the changing distribution of labour between the coalfields and major wage payment categories. This analysis also provides a useful insight into the mechanics of occupational wage differentials.

While our analysis suggests that miners' wage movements have displayed some propensity to wage drift, it also suggests that previous attempts to measure the impact of central wage decisions on wage movements have been rather misleading because they have paid insufficient attention to the mechanics of pay determination and the scope and intention of national agreements; and particularly so in the case of the piecework structure.

The third part takes up three other important issues of concern to wage-policy practitioners. These are (in turn) the impact of wages policy on productivity, disputes and strikes, and labour supply. Each chapter sheds new light on certain generally controversial issues, in a coalmining context. For example, the analysis suggests that the changeover from piece to daywage payment systems at the coalface resulted in many miners revising their notions of a fair day's work. The chapter on strikes suggests that wage structure simplification was of less importance in the industry's declining strike frequency since the 1950s than earlier writers have suggested, and that its most significant effect was to make the miners' organisation itself a more effective force, particularly in its attempt to restore the miners' living standards which were severely eroded (to some extent in absolute terms but certainly in relative ones) during the late 1950s and the

1960s. On labour supply, we examine the sensitivity of labour movements between mining and other industries to changes in relative wages. While the analysis suggests that changes in the miners' external pay relativity have not been wholly incompatible with the employment requirements of the industry, it also emphasises the dangers inherent in a major disturbance of relative wages of a substantial industrial group.

In the fourth part the writer first draws conclusions from the mining case which are relevant to wage-policy practitioners more generally. This is followed by a postscript which examines the impact of the productivity schemes which were introduced towards the end of 1977.

It should be stressed at this stage that this study refers only to industrial workers employed in coal mines operated by the NCB. Although the small numbers of workers employed in the remaining mines operated under private licence from the NCB are subject to the nationalised industry's wage agreements, they are excluded from this analysis, as are those industrial workers employed by the NCB in activities and services ancillary to coal mining. The latter include industrial workers employed in the NCB's coal products division, brickworks, central workshops, and in transport and distribution outside the colliery. By definition all non-industrial staff employed by the NCB are also excluded. In their negotiations, the NCB and the unions concerned have set the non-industrial staff apart from the bulk of the industry's industrial workers; in holidays, sickness benefit, etc., they have a special status.

1. THE ENVIRONMENT FOR WAGE DETERMINATION

Throughout its history, and during the post-war period in particular, the coalmining industry has had to adapt to new situations. And in no area of policy has that need been greater than in that of wages. In this book we demonstrate that wages policy in coalmining since nationalisation reflects a continuous response to the market, technological and institutional environment in which the industry has operated and that only by examining the complex interaction of these forces can we understand the salient features of wage behaviour in the industry during that period. We begin this study, therefore, by acquainting the reader with some of the most important post-war developments in the industry, and some of its principal features, insofar as they have a bearing on wage decisions and resultant wage levels and movements.

1.1 The market environment

Since nationalisation in 1946 the coal industry has experienced dramatic changes in its market environment, some features of which are illustrated in Figure 1.1. In 1946 the industry supplied over 90% of UK primary fuel requirements and had a virtual monopoly of the UK fuel economy: oil was imported in only small quantities and other forms of primary energy, such as natural gas, were unavailable. Furthermore, coal was in very short supply. The needs of war had stretched the mines to the limit and, by the cessation of hostilities, the industry was short of manpower and sorely pressed to meet the added demands of an energy-starved Europe.

Between 1947 and 1956 the demand for coal was expanding within a situation of general energy shortage. At home, coal prices and allocation were regulated by the government, which also took steps to encourage fuel efficiency. Nevertheless, inland coal consumption rose by 33 million tons over this period. Coal output responded to the increased demand until 1953, and home requirements would have been met had not some 15 million tons on average been exported annually. Between 1953 and 1956, however, coal output stagnated and that of deep-mined coal actually fell slightly. Coal stocks, already depleted by the War effort, were insufficient to bridge the gap

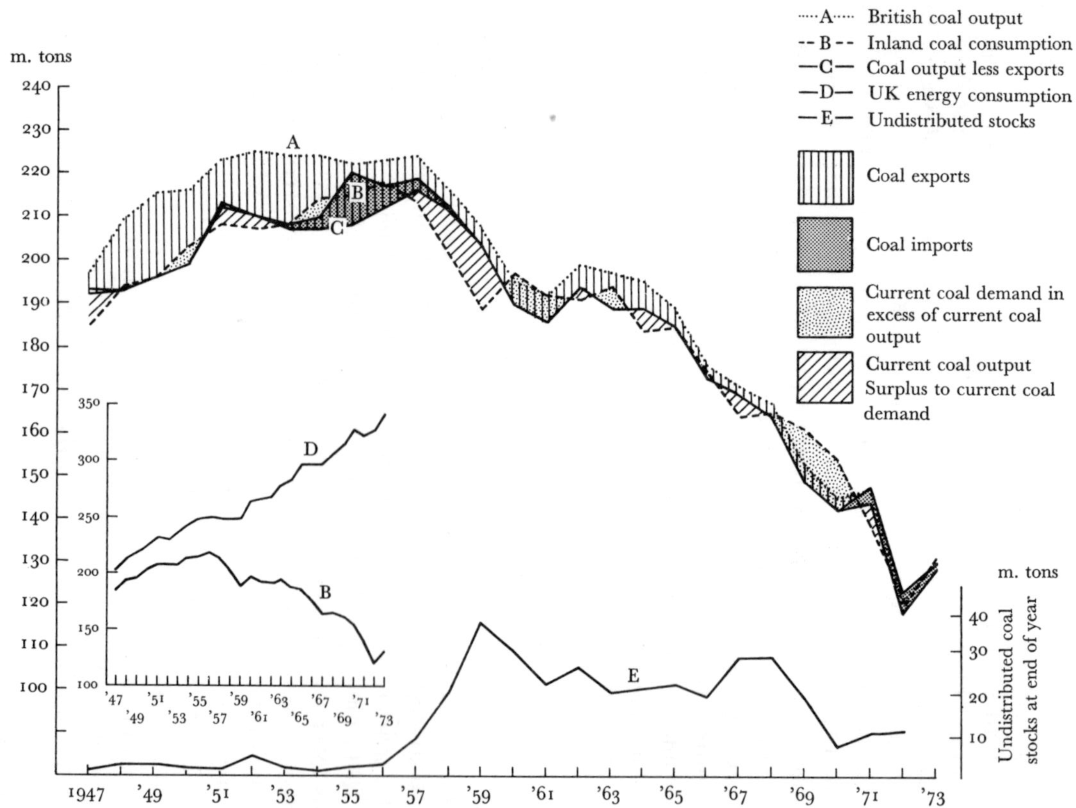

Figure 1.1 The coal and energy pattern 1947–73

between total demand and current supply, and, for the first time in Britain's history, coal had to be imported. This, then, was the period of 'coal at any price'.

The market environment began to change in 1957: coal consumption fell by 8 million tons in that year and by 1959 total demand was 36 million tons below that of 1956. Coal stocks, which had never exceeded 5 million tons per annum prior to 1956, reached the unprecedented height of 36 million tons by 1959.

Although the initial fall in coal demand of the late 1950s coincided with the general recession, it soon became apparent that a radical change had occurred in the energy market. Inland coal consumption revived slightly in 1960 but, whereas the total energy requirement had risen to a new all time high, coal demand was 10% below that of 1956. By 1967, total coal consumption had fallen to 166 million tons – some 36 million tons below the level of 1960 – and, although this represented a slower rate of decline than during the period 1956–59, the downward trend was accelerating in the mid 1960s. In response to this reduced demand, coal output was cut from 195 million tons to 171 million tons between 1960 and 1967 and that of deep-mined coal from 184 million to 164 million tons.

The decline in coal's popularity in the late 1950s and 1960s was attributable almost entirely to the successful competition of substitute fuels, and oil in particular. Although oil consumption had increased from 18 million tons of coal equivalent[1] in 1947 to 38 million tons in 1956 most of this increase had been complementary to the increase in coal consumption. The opening of the Middle Eastern oil fields and their rapid development following the Suez crisis, however, together with the flood of cheap oil into Western countries following the oil import restrictions in the USA, greatly reduced the price of oil relative to coal and thus encouraged the conversion of installations from coal to oil firing. In the space of a few years, then, coal had lost its stranglehold in what had become a highly competitive UK energy market: by 1967 total oil consumption in the UK was running at 119 million tons of coal equivalent.

Although coal's declining popularity in the mid 1950s was due in part to spiralling costs, which had caused the price of coal to increase by 8% per annum during the first decade of nationalisation, the industry was hardly to blame for its predicament. Productivity in the mines had been falling steadily in the decade prior to nationalisation and the industry was in need of a massive injection of capital investment. In 1947 plans were laid for an anticipated output of 240 million tons of coal by the mid 1960s. In the meantime output had to come from existing pits many of which would have been closed on economic grounds in normal circumstances. Because coal is an extractive industry, output can only be maintained in the long run, and costs contained, by reconstructing pits and sinking new ones. But this takes time and meanwhile the cost of producing a given output rises sharply as coal is worked further away from the pit shaft.

In the new market environment it became imperative that the industry reduce, or at least contain, its rapidly rising costs of production. The immediate reaction was to start closing uneconomic collieries which had been kept open during the production drive. In 1957, 27 pits were closed. But in the next ten years the rate of closure accelerated so that by 1967 half the pits which had existed in 1957 had disappeared.

The need to contain costs also gave an added urgency to the technological revolution which was about to besiege mining. Due partly to inertia and partly to difficulties in the experimentation stages, the replacement of machine cutting, which was tied to a cyclic system of working, by power loading whereby coal could be cut and loaded in one mechanical process, had made little headway in the early 1950s. However, between 1956 and 1960, the percentage of coal obtained from power loading increased from 15.5% to 37.5% and by 1968 approximately 90% of total production was obtained from power-loading faces. Nor has mechanisation been confined to the coalface: coal transport, winding, washing, packaging etc. have all been affected.

The effects of this extensive reconstruction are immediately apparent in Figure 1.2. Productivity (output per manshift) increased from 25 to 28 cwts per manshift overall between 1957 and 1960. Thereafter it increased fairly steadily to 42 cwts per manshift overall in 1968. This upward productivity trend naturally had a favourable effect on costs: between 1957 and 1968 costs rose by an average of only 1.6% per annum compared with an average annual increase of 8% in the period 1947–56. The wages proportion of total costs of production fell from 60% to 46% over the same period.

The main brunt of all this upheaval was inevitably borne by the labour force. Having fluctuated between 700,000 and 720,000 between 1947 and 1957, the number of wage earners employed at collieries slumped sharply thereafter and fell to around the 300,000 mark by 1970. Some redundancies were inevitable but by reducing working hours, restricting recruitment, cutting coal imports, phasing colliery closures, placing a heavy reliance on natural wastage and some protection from the government (which imposed a levy on fuel oil and provided substantial subsidies), the worst effects of the run-down on unemployment were mitigated. One side effect of this rapid

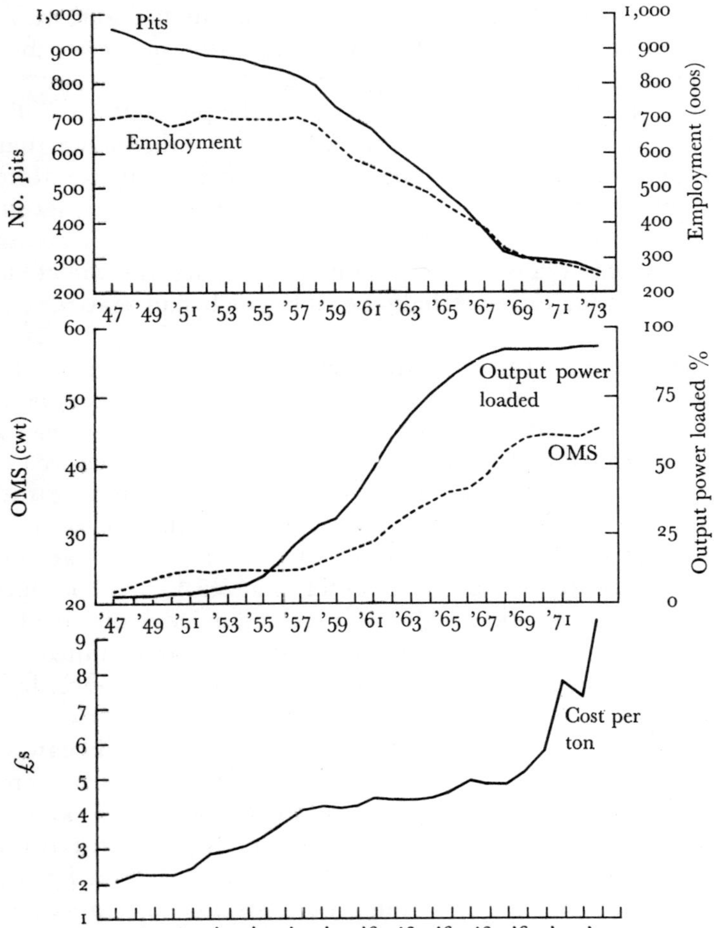

Figure 1.2 Trends in employment, technology, productivity and cost 1947–73
Source: Ministry of Fuel and Power, *Digest of Energy Statistics*

run-down in employment, however, was a considerable ageing of the industry's labour force: the proportion under 40 years of age fell from 50% in 1958 to 36% in 1971.[2]

The late 1960s represent the beginning of another phase in coal's post-war history. On paper at least, coal seemed to be entering an even more competitive phase than hitherto, with the development of both nuclear energy and the discovery of North Sea gas. This led the government, in its White Paper on Fuel Policy[3] of November 1967, to estimate a coal demand of 120 million tons in 1975. Accordingly, the contraction of the industry's capacity continued: in the accounting year 1968–69 some 59 collieries were closed or merged and the labour force decreased by 12.6%, which was the highest rate of run-down since nationalisation.

In fact, though, coal consumption first rose slightly and then levelled off in the late 1960s, with the effect that current demand exceeded current supply, the deficiency being made up from undistributed stocks which, with the added depletion on account of industrial action, fell from 28 million tons in 1967 to 7 million tons in 1970 when coal was once again imported in small quantities. The supply situation was aggravated further by the national strike of 1972, by which time it was also evident that some fairly fundamental changes were occurring in the world energy market. First, the continued expansion of world oil consumption required more extensive and costly exploration. Second, it was obvious that the oil producers were going to demand a larger share of oil revenues. Third, the USA was about to become more dependent on external fuel suppliers. Of these factors the quadrupling of imported oil prices in less than a year had the most important implication for coal in the short run: instead of facing a falling market for its product the industry was looking forward to an expanding market.

Despite the emerging trends, the industry continued to contract until 1973 by which time employment had reached 248,000. By then, however, the problem was one of manpower shortage. Furthermore, the pace of technological change was slowing down. These factors, combined with serious industrial unrest, had the effect of considerably slowing down the rate of productivity growth and making even planned output targets difficult to achieve. In 1975 the estimated demand for coal in the late 1970s was tentatively put at 150 million tons. The industry, on the other hand, was programmed for an output of 120 million tons which, even having stabilised manpower, it would have done well to achieve.

At this point it is well to appreciate that, in almost every respect, coalmining is still a diverse industry and that some coalfields have been better equipped than others to withstand the pressures outlined above. Coal itself is a heterogeneous product[4] and, while several varieties are usually found in each coalfield, the distribution of the various types is far from even: South Wales has a virtual monopoly of anthracite and is famous for its high quality steam coals; coking and gas coals come primarily from Durham, although these can also be found in South Wales, Kent and North West Yorkshire, while the Midland areas and Scotland produce the best house and general purpose coal.

The diversity of the product is matched by the variations in geological conditions to be found in the industry, which almost defy generalisation. True, the coal seams of Yorkshire tend to be deeper than elsewhere, and seams tend to be thin in Scotland and Durham and thick in the East Midlands; but within every coalfield there are large variations in these and in other aspects of geology, such as seam gradients and faulting, etc.

These very real differences in the product and its markets and mining conditions have inevitably made for very considerable differences in working results, of which Table 1.1 provides an illustration for the years 1963–66.

The table shows that the more highly mechanised areas of the East Midlands, Yorkshire and the West Midlands, selling the relatively low-priced general purpose coal for electricity generation upon which the industry has become so dependent, had a clear advantage over the other Divisions in both productivity and profitability. Again, though, these average figures hide enormous local variations. Thus, while the Welsh coalfield shows low returns, these are almost wholly due to poor results in West Wales. Likewise the Western and older part of the coalfield adversely affects the general returns in Northumberland and Durham.

As the industry became increasingly dependent on the demand from power stations it was inevitable that output and employment should become increasingly concentrated in the low-cost coalfields of the Midlands and Yorkshire. In Table 1.2, for example, we can see that, although pit closures and reduced employment were a feature of

Table 1.1. *Some aspects of regional diversity (annual averages 1963–66)*

Coal board division	Output per manshift	Cost per ton (shillings)	Proceeds per ton[a] (shillings)	Percentage of coal power loaded
Scottish	30	106	105	69
Northumberland and Durham	30	99	98	64
Yorkshire	38	85	89	81
North Western	31	110	106	90
East Midlands	50	72	80	90
West Midlands	40	91	95	80
South Western	25	119	116	57
Kent	29	112	102	83

[a] Before charging interest.
Source: NCB, *Report and Accounts* 1963–66.

Table 1.2. *Regional trends in employment and output*

	Number of producing collieries				Employment (000s)				Output (m.tons)			
	1947		1974[a]		1947		1974[a]		1947		1974[a]	
	No.	%	No.	%	No.	%	No.	%	No.	%	No.	%
Scottish	198	21	22	9	81	11	24	10	22	12	10	9
Northumberland and Durham	188	20	34	14	150	21	38	15	35	19	15	13
Yorkshire	118	12	65	26	139	20	66	27	38	21	32	28
North Western	85	9	12	5	66	9	13	5	14	8	5	3
East Midlands	99	10	66	27	95	13	71	29	35	19	43	37
West Midlands	65	6			60	8			17	9		
South Western	201	21	44	18	116	17	31	13	22	12	9	8
Kent	4	1	3	1	6	1	3	1	1	1	1	1
Total	958	100	246	100	711	100	246	100	184	100	115	100

[a] The data for 1974 refer to the new administrative areas of the NCB. They have been regrouped to conform to the old divisions.

Source: NCB, *Reports and Accounts*, 1947. Department of Energy, *Digest of Energy Statistics*, 1975.

every coalfield, the labour forces of the East Midlands and Yorkshire coalfields fell much less than those of other areas, as did output, which was relatively stable, while in others it more than halved. Despite the more favourable market environment of the 1970s, the trend towards concentrating coal production in the relatively low-cost coalfields of the Midlands and Yorkshire will almost certainly continue.

1.2 Work organisation, occupations and payment methods

Techniques of production will determine the skill composition of an industry's labour force and may also help to influence the method of wage payment applied in particular cases. In this section, therefore, we outline the basic principles of mining coal and describe some of the major changes which have occurred in colliery working methods since nationalisation.

1.2.1 *Mining methods and occupations*

Deep-mined coal is obtained by sinking shafts from the earth's surface and driving headings, or tunnels, into the coal seam. Tunnels are driven 'in bye' from the shaft bottom in sets of three. One tunnel is used as a travelling road for miners, another for haulage and another for foul air extraction.

The oldest system for extracting coal is known as the 'room (or bord) and pillar'. Under this system a series of narrow headings are driven into the coal seam at right angles from the main road. These are then connected by cross-headings to form pillars of coal which are subsequently extracted. Although relics of this system still exist in the North East, it accounts for an insignificant proportion of coal production today.

When the mining industry was nationalised, 87% of deep-mined coal was obtained from longwall methods, and today 97% is so obtained. By this system, coal is worked from a single face, which is usually between 150 and 300 yards long, running at right angles between two main roadways. In any one pit several faces may be worked simultaneously by different groups of men. As coal is taken in strips from the length of the face, so the face advances into the coal seam, and the roof in the vicinity of the face is supported. This leaves a space behind the coalface (the 'gob'). This used to be packed with waste material but now the roof is allowed to collapse into it. In order that materials can get to the face, and the coal from it, the roadways at either end must keep in touch with the working area. The roadways have to be considerably higher than the working face, which usually means cutting into a stone roof. The rock from the roof is either used to pack the roadway walls or is thrown into the 'gob' behind the face. Once the roadway has been hewn out it is supported with steel arches.

The work of advancing the roadways has not changed much over the years, and is performed by small groups of four or five men, known as 'rippers'. Their task consists of drilling and blasting the roof, packing the waste and arching the roadway. Their work is generally recognised to be among the most arduous and skilled in the pit. The need to keep the roadway in touch with the face means that the rippers are under constant pressure to work at speed.

The methods of working at the coalface have changed markedly since nationalisation. By 1946 one could still find relics of the traditional *hand-got system* of mining, whereby a string of colliers working along the face would each hew coal, fill it into tubs alongside them, rip the ceiling and pack and support their respective working places – but for the most part it was confined to bord and pillar workings. Elsewhere, this method had given way to *machine mining*. This, however, was mechanisation in its most infant form. True, the coal was under-cut by a machine, but it still had to be blasted down from the face and all the ancillary operations previously performed by the collier remained an essential part of the production process. Furthermore, with the advent of machine mining, the various aspects of the collier's job became specialist tasks, which had to be carried out in a definite cycle. Thus, the coal would usually be under-cut on the night shift. On the morning shift, the coal would be blasted down and props set to prevent the roof from collapsing, and the coal laboriously filled by hand on to conveyors running alongside the face. On the afternoon shift, the conveyor would be dismantled and re-erected nearer the face by 'conveyor movers', packs built (by 'packers'), to control the collapse of the roof, and the supports removed from the line of earlier workings to allow the roof to subside. This cycle of operations would then recommence from the night shift.

The next change in working methods at the face came with the advent of the *power loading* machine, which combined coal cutting and loading in one single mechanical operation. The machine runs on rails and is pulled along by a chain which runs alongside the face. As the coal is dislodged it is thrown, by the machine, on to a belt, which transports the coal along the face to the main roadway and thence to the pit bottom.

The power-loading machine is operated by one man, whose job it is to ensure that the machine maintains a straight path along the face. He is supported, however, by a team of men engaged in more arduous operations. Before the machine can make a fresh cut it has to be moved forward so that the cutting equipment is behind the coal to be removed. Thus, room has to be cleared at the face end to accommodate the machine at the end of each cut. This area is known as the stable hole. Some machines will cut either way, in which case two stable holes (one at either end of the face) have to be provided. The conventional equipment for making the stable holes has long been the pneumatic pick, dynamite, shovel and physical effort.[5]

In its early days, power loading was also tied to a cyclic system of working. Thus, before a new cut could be made, the conveyor belt had to be dismantled and brought up to the face, as did the supports which were necessary for controlling the roof. These tasks would involve up to 15 men on a 150 yard face. The breakaway from the cyclic system to continuous working was made with the advent of flexible conveyors, which could be 'snaked' forward to the face as soon as the machine had passed. This development also facilitated the adoption, in the early 1960s, of 'self-advancing powered supports'. These are steel props with large supporting roofs set on a firm protruding base, which are operated hydraulically. As air is let out of a tap, the support moves forward, pushing the conveyor into position at the coal face. When the props are set apart at two foot intervals along the face, the roof immediately above the working face is supported by a solid steel sheet. As the props move forward, so the roof behind is left to collapse into the gob. These technical advances have necessitated that much attention be given to maintenance, and to cover this, craftsmen now work in conjunction with the power-loading teams.

This description of working methods has covered most of the main occupations found at the coalface, and their changing character. They are not, however, the only jobs which are performed underground. Craftsmen are also employed to look after the electrical system upon which the modern pits are so dependent. The other important jobs elsewhere underground (i.e. between the pit bottom and the coalface) are concerned with transport and maintenance of the roadways. Roadway repairs are conducted by groups of men known as 'back rippers', who do the same type of job as the face rippers. They are also responsible for opening up new faces. Another important task, of course, is that of supervision.

When the coal reaches the pit bottom it is loaded into a bunker. There it is fed into skips and taken up the shaft to the surface. The winding gear which is used to get men down the pit, as well as to haul coal up (but not at the same time), is operated by the winding engineman. Apart from this, surface labour is employed in general labouring, cleaning and despatching the coal, machine repairs and supervision.

The distribution of the colliery labour force between broad occupational categories at the coalface, elsewhere underground, and at the pit surface, is shown in Table 1.3. In 1952, approximately 78% of colliery industrial workers were employed underground, and this percentage has shown only slight variation since then. However, the proportion of faceworkers in the total labour force has been reduced by six percentage points, while the percentage of manual workers elsewhere underground has increased by three points, the difference being made up by a proportional increase in the number of non-manual supervisory staff. This redistribution of the underground labour force partly reflects changes in the statistical definitions of 'face' and 'elsewhere underground tasks'. It also reflects the impact of power loading, which has served to reduce the labour required at the coal face in both absolute and relative terms, and, because of the greater speed of face advance, has also necessitated a greater proportion of men being employed on development and repair work in the roadways – these being crucial to an efficient supply system to the face – and on maintenance generally. At the surface more efficient methods of preparing coal, automatic pit winding, and the increased

Table 1.3. *Occupational classification of wage earners on colliery books at NCB mines*

Occupation at main places of work	1952 (000s)	1952 % of total	1967 (Dec.)[a] (000s)	1967 (Dec.)[a] % of total
Underground officials				
Overmen	4.5	0.7	5.0	1.3
Deputies	20.4	3.1	17.6	4.6
Shotfirers	8.7	1.3	4.4	1.2
Other underground officials	2.6	0.4	3.8	1.0
Face workers	263.4	39.9	131.3	34.2
Workers elsewhere underground				
Development at roadways	16.6	2.5	14.2	3.7
Transport	107.6	16.2	54.2	14.2
Other underground labour (including roadway repairs, engineering services and reconstruction)	94.0	14.1	71.2	18.6
Total	517.9	77.9	301.7	78.8
Surface workers				
Weekly paid industrial staff	4.5	0.7	4.2	1.0
Winding and banking	19.4	2.9	8.9	2.3
Traffic	11.9	1.8	10.2	2.7
Coal preparation	29.5	4.4	13.9	3.6
Power	9.2	1.4	3.1	0.8
Maintenance	36.7	5.5	20.7	5.4
Other surface labour	35.7	5.4	19.7	5.2
Total	146.9	22.1	80.6	21.0
Grand total	664.8	100.0	382.4	100.0

[a]This table was discontinued in 1968 and replaced by employment estimates based on shifts worked.

Source: Ministry of Fuel and Power, *Statistical Digest*, 1953 and 1960 and National Coal Board, 1967.

emphasis on supervision have served again to bring about small shifts in the distribution of labour between tasks.

It should not be assumed, of course, that the distributions shown in this table are by any means representative of particular pits. Underground mining conditions vary so greatly that employment has to be tailored to specific requirements. In some pits, actual geological conditions, for example, may dictate that more men be employed at the face than may be thought ‘normal’. Similarly, a manager who is anxious to open a new face may have a disproportionate allocation of labour on development work.

1.2.2 *Methods of payment*

For the most part, surface work has always been paid for by the shift or day, while that of supervision has been paid for on a fixed weekly basis. The daywage principle was also applied to most jobs elsewhere underground, with the exception of back ripping. The coal face has traditionally been the stronghold of piecework. In 1966, however, a start was made in phasing-out piecework on coal face operations in favour of a fixed daywage payment system. The daywage principle was also extended to piecework tasks elsewhere underground in 1971.

1.3 Management organisation

The most outstanding feature of the privately owned coalmining industry's organisation was its extreme fragmentation: despite the amalgamation of colliery companies during the 1930s, the industry's assets were still controlled by some 740 undertakings in 1944. The companies differed markedly in size, from the huge Powell Duffryn complex, which owned over 93 pits and employed 38,000 men, to small undertakings owning one pit and employing only a few men. Although the employers' general interests were represented by district employers' associations, affiliated to the Mining Association of Great Britain, cooperation between the colliery companies was minimal. The deleterious effects of the industry's fiercely competitive instinct, and fragmentation, on investment and general planning were highlighted in 1945 by the Reid Committee,[6] whose emphasis on the need for larger operating units, and a more co-operative approach, was taken by the then Labour Government as a powerful argument in favour of nationalisation.

The coal industry nationalisation act of 1946 gave the newly constituted National Coal Board (NCB) the task of devising an organisation to maximise the advantages of planned production. The 958 pits which the NCB inherited were subsequently grouped into 48 Areas, each under the charge of a General Manager. These areas, in turn, were grouped into nine Coalfield Divisions, which followed the regional groupings established by the Ministry of Fuel and Power during the War. Apart from small changes in Area and Divisional boundaries, this pattern of organisation remained intact until March 1967, when the divisional structure was abolished and the old areas reorganised into 17 'new style' Areas.

The NCB's management structure followed the 'line and staff' principle. Thus, whereas pit managers were in charge of daily work schedules, their general directives came from the Area General Manager who was in turn responsible to the Divisional Coal Board, which acted within the broad constraint of national policy as laid down by the NCB in London and, above that, by the Minister of Fuel and Power. The 'staffs' were departments at national, divisional and area levels to assist in problems of production, finance and labour. This principle was retained in the 1967 reorganisation, the only difference being that the 'new style' Areas are now directly responsible to London.

At nationalisation, wages constituted some 65% of total costs and their importance in cost control should have been obvious to anyone. Despite this, the Fleck Committee[7] found the administrative organisation for wage control hopelessly lacking and that for industrial relations generally too diffuse, spread as it was between three departments. This criticism led to the NCB establishing a new industrial relations department within which there was to be a new sub-department named 'wages and control', with special responsibility for monitoring wage agreements as well as conducting negotiations with the trade unions on wages and conditions. The impact of these changes in organisation and wage levels will be of particular concern to us in later chapters.

Despite the elaborate administrative hierarchy of the NCB, the colliery is still the basic unit of operational management. The colliery manager is responsible for the day-to-day running of the pit, and it is his job to ensure that output targets are met. Nevertheless the colliery manager's job has always had a strong personnel element in it. Administration requires that much of the pit manager's time is spent in the office rather than down the pit. Underground operations are usually in the charge of an under-manager, who is assisted in turn by groups of overmen, with deputies under them. Historically, the deputies have been in charge of pit safety but, with increased mechanisation, they have tended to assume added managerial responsibilities. The overmen and deputies are in direct contact with the men, and it is with them that much of the day-to-day difficulties are first discussed and, as first line management, they have traditionally played an important role in the wage control function.

There can be little doubt that the overall powers of pit management have been severely curtailed since nationalisation. Production and cost targets, for example, are set for each pit by area management, and the pit manager has little say in the production methods used or in colliery development schemes. With the gradual extension of national wage agreements since nationalisation, the scope for local manoeuvring on pay questions has also been reduced. Pit management has undoubtedly become more professional since nationalisation. In the very largest pits (over 750,000 tons per annum), many of the manager's traditional responsibilities have been hived off to deputy managers and it is not uncommon also for the pit manager to be assisted by a personnel officer. While these developments are partly the consequence of the growing size of pits, they are generally indicative of the growing sophistication of the managerial function.

1.4 Trade unions

Coalmining has a long history of collective organisations. Before 1944 the majority of colliery workers were affiliated, through local and district associations and federations, to the Miners' Federation of Great Britain (MFGB) formed in 1886. Trade union organisation was along industrial lines, in the sense that the MFGB admitted only mineworkers. Craftsmen, however, tended to belong to separate craft unions which were affiliated to national federations catering for their particular occupations.

While the MFGB represented the bulk of miners on many important labour issues, it had virtually no power to intervene in wage issues. Collective wage fixing in the privately-owned industry was entirely a local and district affair and the sole prerogative of the district miners' associations, all of which jealously preserved their autonomy within the federation. Disillusioned by the industrial defeats of the 1920s and 1930s, and in preparation for an extension of national bargaining following the Report of the Greene Tribunal in 1943, however, the miners finally effected, in 1945, a much discussed reform of their organisation, which resulted in the creation of a National Union of Mineworkers (NUM). The NUM subsequently secured the sole rights to negotiate with the NCB on behalf of colliery manual workers.

The administrative organisation of the NUM is based on twenty Area Unions, fourteen of which are regional in character and six occupational. The latter Areas consist of two groupings of craftsmen (who joined the NUM in 1945), two groups whose members hold dual affiliations with the NUM and one or other of the two large general unions (the Transport and General Workers (TGWU) and the General and Municipal Workers (GMWU)), the COSA, representing white collar workers, colliery specialists and engineering supervisors, and a group of cokemen.

The local basis of trade unionism in the mining industry is the lodge, centred on the place of work. The lodge committee is usually composed of men from each of the main working areas of the pit. The lodge secretary and president are usually part-time, but in some of the larger pits in the East Midlands and Yorkshire the secretary is full-time. Much of the lodge's importance has derived traditionally from its key role in local piecerate negotiations which helps to explain why in the past the positions of secretary and president were usually occupied by faceworkers. Today, however, it is not unusual to find craftsmen occupying the secretary's position.

While nationalisation has enabled the unification of management to proceed steadily, the problems on the trade union side of the industry have been far greater for, despite its name, the NUM is still very much a federation of relatively autonomous unions. The regional areas of the NUM, for example, are none other than the old District Miners' Associations, some of which are themselves federations of local unions. Again, the craft groupings are made up of local and national craft unions which, like the District Miners' Associations (or areas) and their affiliated organisations, have retained their individual identity within the NUM. Thus, each of the Constituent Associations of the NUM has retained its right to register as a separate trade union, elect its own officials, set its own dues, operate its own benefit programmes, hold its own annual conference and, in the case of some of the craft associations, maintain lodges which are quite separate from those of the regional area unions.

The main change brought about by the transformation of the MFGB to the NUM was the obligation it imposed on the Area Unions to abide by the decisions of the Union's National Executive Committee on wage questions. Thus, the Area officials became responsible to the National Executive Committee for 'all negotiations for price lists, wages, compensation and all other matters which purely relate to a colliery and district'[8] and in 'the case of negotiations in connection with the arrangement of area and district agreements no area shall ... complete any such negotiations and enter into an agreement without the previous approval of or under power delegated to them by the National Executive Committee'.[9] Also, 'The National Executive Committee, if it thinks advisable, may withhold its approval of any proposed agreement with a view to any outstanding points of difference being recognised or otherwise.'[10]

Although these rulings appear to give the National Union considerable control over its constituent parts, they have seldom been invoked and in local and district wage questions the Areas retained much of the autonomy they enjoyed in the days of the MFGB. Such power as the National Union has obtained over its constituent parts, on wage questions, has really been the consequence of its negotiating national agreements to replace those of a local and district character. Even in the formulation of national wages policy, however, the influence of the Area Unions is very much in evidence, for the National Executive Committee is composed entirely of officers elected from the Area

Unions, the latter's representation being proportional to Area membership, with the proviso that each area shall have at least one representative.

The changing regional locus of coalmining has caused a considerable shift in the balance of power between the areas of the NUM, as has the changing technological basis of the industry. Table 1.4, for example, shows that the regional areas of the Union have suffered a

Table 1.4. *NUM: Area membership, delegate representation at annual conference and representatives on National Executive Committee*

Geographical areas	Membership %		Delegates at annual conference %		Number of representatives on the National Executive Committee	
	1953	1975	1953	1975	1953	1975
Cumberland	0.8	0.4	1.4	2.9	1	1
Derbyshire	5.0	4.3	4.9	4.3	1	1
Durham	11.2	7.5	10.4	5.7	2	2
Kent	0.9	0.9	1.4	2.9	1	1
Lancashire	6.5	3.6	6.3	2.9	1	1
Leicestershire	1.0	1.0	1.4	2.9	1	1
Midlands	7.0	5.4	6.9	4.3	1	1
Northumberland	5.2	3.2	4.9	2.9	1	1
N. Wales	1.1	0.4	1.4	2.9	1	1
Nottingham	6.1	12.0	6.3	10.0	1	2
Scotland	10.0	7.4	9.7	5.7	2	1
Somerset	0.4	*	1.4	*	1	*
S. Derbyshire	0.7	0.9	1.4	2.9	1	1
South Wales	14.2	10.7	13.2	8.6	2	2
Yorkshire	17.0	24.4	13.9	18.6	2	3
Total	87.1	82.4	84.9	77.0	19	19
Occupational 'Areas'						
Cokemen	1.0	2.0	1.4	2.9	1	1
Group No. 1	3.0	4.1	3.4	5.7	1	1
Group No. 2	2.4	2.0	2.7	2.9	1	1
COSA	3.5	6.8	3.5	5.7	1	1
Power Group 1	1.7	1.9	2.7	2.9	1	1
Power Group 2	1.3	0.8	1.4	2.9	1	1
Total	12.9	17.6	15.1	23.0	6	6

Note: The percentage columns may not add to 100, due to rounding. * Somerset merged with South Wales.

Source: Report of NUM Credentials Committee to Annual Conference in NUM *Annual Conference*, Report and Proceedings 1953 and 1975.

proportionately greater loss of membership than the occupational areas and that their voting representation at the National Conference (the main decision making body of the Union) has declined relative to that of the occupational areas. Although the trend in favour of the occupational areas is not dramatic, it has to be reckoned with in a narrow vote, as does the declining voting power of the traditionally 'left wing' areas, such as South Wales, Scotland, Derbyshire and Kent. In this context, however, the recent swing towards the 'left' in Yorkshire is very important. Within the Union the more meaningful distinction on wage-policy questions has often been between what were the high- and low-paid areas. If we now group areas, such as Lancashire, and Durham, with South Wales, Scotland (also traditionally lower-paid areas), Derbyshire and Kent, then the balance of power can certainly be seen to have moved against what were the lowest-paid areas.

Although we have tended to stress the importance of area loyalties in the NUM, it must be appreciated that the very diversity of its membership has often militated against unity. It is very seldom that occupational interests coincide. When they do it is usually when the labour force is threatened by some common external force. Once one realises this fact, it is clear that any reference to a 'union wages policy', must be treated with due caution.

Although pride of place in our subsequent examination of union wages policy must inevitably go to the NUM, not all the work people with whom we are concerned in this study are represented by that union. Of the exceptions the underofficials are the most important group. They have their own organisation and, while in some coalfields (notably Durham, Scotland and Nottinghamshire) a small proportion of overmen and deputies are members of the NUM, they are represented in the main by the National Association of Colliery Overmen, Deputies and Shotfirers (NACODS). To avoid the difficulties which mixed affiliations in some coalfields might cause, however, the NCB agreed at nationalisation to negotiate over overmen and deputies with the NUM and NACODS in a joint capacity. A similar arrangement existed in respect of weekly paid industrial staff (WPIS), some of whom were members of NACODS while the majority were members of the NUM. 10a

1.5 Collective bargaining machinery

The most outstanding feature of the nationalised coalmining industry's arrangements for collective wage regulation has been the provision of formal procedures for collective bargaining at three levels, with arbitration as the final stage in each case should there be a failure to agree. At national level, wage questions are dealt with by the Joint National Negotiating Committee (JNNC), which consists of sixteen representatives of the NCB and an equal number of NUM representatives. Should the JNNC fail to agree a settlement within five weeks, a question can, if neither side objects, be referred for arbitration to the National Reference Tribunal.

For the purposes of wage regulation, the industry has been divided into 17 wage districts.[11] These date back to pre-nationalisation days and have boundaries which are confusingly different from those of the NUM Areas, which are, of course, different again from those of the old and new NCB Areas. Within each wage district there is a Conciliation Board, composed before 1966 of representatives of the Divisional Coal Board (but now representatives of the Area Boards) and of the relevant Area Unions and other Constituent Associations of the NUM. If the District Board fail to settle an issue within three weeks, the parties can agree to refer it to a district referee. When a district question is considered to have national implications, however, it may be referred to the JNNC for settlement.

Although there had long been provision for collective bargaining at district level, the national machinery and the detailed form of the district machinery described above date from 1943, and followed the recommendations of the Greene Tribunal a year earlier. In addition there has been a formal procedure for wage regulation at pit level since 1947. At the pit, formal negotiations are conducted between the secretary of the miners' lodge and the pit manager. If they fail to agree, the question is then referred to a District Disputes Committee and, thereafter, to an Umpire taken from a panel agreed by the NUM and NCB. Should an issue arise at pit level which is of district or national importance, it may be taken out of the pit procedure and referred to the District Conciliation Board or the JNNC, as appropriate.

The logic of these arrangements for collective wage regulation, as well as of developments in the industry's wage structure since nationalisation, is best understood in relation to the evolution of the industry's methods of wage fixing before nationalisation. Let us turn, then, to consider these briefly.

1.6 The evolution of wage regulation in the coalmining industry before nationalisation

Before the Second World War a miner's wage was composed of two principal elements – a basic rate and a percentage addition. The basic, time or piece rate was negotiated locally and intended to reflect differences in work content and the skills of workers. Far from being uniform in a locality, however, the basic rate for similar work varied from pit to pit, and seam to seam, and even from one part of a seam to another, according to physical conditions and the employer's ability to pay. Basic rates were sometimes varied with changes in working methods and physical conditions; but such changes were infrequent and, once fixed, the basic rate was seldom altered during the life of the seam or pit.

The other wage component was the percentage addition to basic rates. This was determined on a (wage) district basis, apart from the period of government control from 1916 to 1921 when wages were regulated by uniform advances or reductions applied on a national basis, and the period 1921 to 1925 when a national wage agreement operated. The percentage addition varied between districts, but, unlike the basic rates, it was uniform between pits in any one area. The value of the percentage addition, however, fluctuated over time, for it was designed specifically to facilitate a speedy adjustment of wage costs to changes in the district's fortunes and thus make the workmen share in the adversity, as well as in the prosperity, of the industry. Thus, one of the main problems inherent in the percentage addition was, as the Samuel Commission so aptly described it, that 'of arriving at an objective test of prosperity or adversity, so as to avoid blind bargaining and trials of strength'.[12]

The percentage addition was regulated automatically, until the late nineteenth century, by reference to sliding scales based on the selling price of coal, or a representative grade of coal, taken at the pithead or port of exportation. Trade union growth owed much to the arbitrariness of the sliding scale system and by the 1890s the percentage addition in most coalfields was determined by conciliation boards,[13] consisting of equal numbers of employer and union representatives presided over by an arbitrator.

Wage determinations through the conciliation boards were probably less arbitrary than the earlier sliding adjustments, if only because they could consider factors other than the selling price of coal when making awards. There was also a tacit agreement that wages should only fluctuate within agreed maximum and minimum limits.[14] However, the miners' leaders, aware of the dangers of sectional bargaining to their own organisations, had become increasingly adamant by the turn of the century that more uniform principles should govern wages. In 1909 the district unions, which had previously opted to stay outside the MFGB, joined its ranks, and in 1910 the MFGB was instructed to replace the district conciliation boards with a national conciliation board, which would regulate wages along common principles on an industry-wide basis. The idea of uniform wage standards – or even principles – was complete anathema to the coal owners, who had always prophesied doom should wages cease to be fashioned by the particular needs and circumstances of individual pits and localities, and they adamantly refused to discuss the issue. The MFGB responded by instructing its constituent associations to discontinue district agreements after 1915.

The advent of war in 1914 gave a sharp twist to the already serious price inflation, which the miners demanded should be offset by a national wage adjustment. Aware that such a concession would help to centralise the MFGB as the accepted unit for collective bargaining, the employers initially refused to negotiate. They eventually agreed to an increase under government pressure but insisted that the allocation be determined in the districts. The trouble following this award[15] (as well as the issue of national negotiations) subsided, however, with the advent of government control of the industry in 1917.

From September 1917, until the end of control in 1921, the district percentages of wages were frozen and wages adjusted nationally by flat-rate additions related to changes in the cost-of-living. This was a great step forward for the MFGB, but it was no long term solution, as the owners' objections to national wage determinations could only be met if the industry remained under unified control and a central profits pool were maintained. After much bitter strife, during which time the government advanced the date for decontrol, the problem was solved in July 1921.

The 1921 Agreement divided the industry into 13 wage districts which, with the exception of the federated area (which was divided into four separate districts), corresponded closely to the span of the old conciliation boards. A conciliation board consisting of equal numbers of workers' and employers' representatives was established in each district. A national board consisting of an equal number of representatives elected by the Mining Association and the Miners' Federation was also established. The respective duties of these boards was not defined beyond allocating the actual settling of wages to the district boards and not, as the miners wished, the national board, which could only arbitrate on disputed district decisions.

Even so, wages were to be regulated by a *uniform* percentage on basic rates according to the proceeds of the district as ascertained by the owners and independent accountants. The wage allocation was 83% of the surplus of the proceeds after deducting the cost of standard wages, other costs of production and profits of 17%. If ascertained proceeds, after the deduction of costs, were insufficient to meet standard profits, the deficiency was carried forward and deducted as the first charge against future surpluses. The agreement also stipulated a minimum fall-back wage in each district, so that no wage was to fall below 20% of the standard wage of the district.[16]

Against the background of increased coal exports arising from the Ruhr occupation, and rising profits, the miners negotiated a new agreement in 1924 which provided for improved wage terms. By the end of 1924, however, coal prices had slumped and, apart from the Eastern District, wages fell to the minimum. The industry incurred a loss during the first half of 1925 and in July the owners, anxious to reduce wage costs and increase the profit share, gave notice to terminate the agreement. The miners' rejection of the owners' proposals led to the appointment of a Royal Commission under the chairmanship of the Rt. Hon. Sir Herbert Samuel, and, while it awaited its outcome, the government decided to bridge the gap between wages payable under the 1924 agreement and those implied by the owners' proposals, with a subsidy which was to last for nine months from 1 August 1925.

Although the Commission supported national wage agreements, it also recommended negotiated wage cuts which would include the minimum wage.[17] The Commission succeeded in pleasing no one and, for the most part, was ignored. The course of the 'general strike' which followed and its results for the terms upon which the mines should be operated after de-control in August 1926 are well known: when the men returned to work in November, negotiations were opened with the owners on a district basis and all semblance of national agreements disappeared.

Although the MFGB was powerless to dictate the terms of district bargaining, its recommendation that its district associations follow the general principles of the 1924 agreement was adhered to in most cases. Hours of work were increased generally,[18] the share of profits from proceeds was increased and there were wage cuts in every district. The wage adjustments were of a similar proportion, however, which helped to contain sectional bargaining. The owners gave assurances that they would not raise coal prices without consulting the government, but this guarantee threw the district wage agreements, based on ascertained proceeds, out of gear and necessitated national wage adjustments, as did the sharp rise in living costs to which miners everywhere were subjected.

During the early years of the Second World War, wages were adjusted by ad hoc flat rate cost-of-living increases; the most important of these being the 1940 'War Additions to Wages Agreement' which, although it left the district wage arrangements intact, formalised the precedent that wages should keep abreast of living costs and that such adjustments should be applied nationally. It was now left to the miners to consolidate their position.

Despite the 1926 debacle, the miners' leaders remained determined to obtain national wage standards. Their campaign gathered momentum again in the late 1930s, and in 1940 they submitted a claim for a national weekly minimum wage for all adult mineworkers and an all-round wage increase. In 1942 the government, aware of dissatisfaction in the coalfields, and its potential effect on output, eventually referred the claim, which had been upwardly revised in the light of wage trends in munitions, to an investigating body under the chairmanship of Lord Greene.

The acceptance of the Greene Tribunal's recommendations represents a landmark in the history of mining wage regulation. Wages were raised by 2/6d per shift and the current district ascertainments were frozen, as minima not maxima. More importantly, though, the miners' wages were to be underpinned by a minimum wage paid on a weekly, and not, as previously, a daily basis, and determined on a national, rather than a district basis. The minimum was first fixed at £4.15 for underground workers and £3.80 for surface workers, thus establishing a differential for underground work. In addition the Greene Tribunal supported the government's own ideas, contained in a White Paper of 1942,[19] 'that a system should be developed by which questions of wages and conditions in the mining industry should be dealt with on a national basis and by a properly constituted national body'.

The national bargaining machinery which resulted from the Greene Tribunal's Report, outlined in section 1.5 above, was soon tested by a substantial pay package in early 1943, which was referred to the National Reference Tribunal headed by Lord Porter. The Tribunal's recommended increase in minimum wages fell short of the miners' claim, however, and disturbed wage differentials (as did its refusal of an application to increase piece rates). Although the owners agreed subsequently to remove these anomalies, provided that the cost of doing so would be met out of the Coal Charges Account, which had been established to permit cross subsidisation of profitable and unprofitable pits, the government refused to meet the cost of price list revisions, despite the fact that many had already been negotiated locally.

The problems arising from the Porter Award highlighted the need for a more fundamental approach to the industry's wage structure. Quite apart from distorting wage differentials, the succession of flat-rate awards seriously complicated wage calculations and, in the case of pieceworkers, had reduced incentives by substantially decreasing the proportion of earnings which were dependent on physical effort. With these problems in mind, the government suggested that the national representatives of employers and miners consider an overhaul of the industry's wage structure.

The subsequent discussions produced the 'Revision of Wages Agreement' of April 1944. To simplify wage compilation, all flat-rate additions to wages, except those of 1940, were consolidated, along with the percentage addition and other local percentages, into the base rate. To restore the pieceworker/daywageman wage differential, however, piece rates were increased by the percentage which the flat rates, excluding the 1940 cost-of-living rate, bore to the appropriate district minimum shift rate. Although piece rates could be revised subsequently in the light of changed working methods, they, and daywage rates, were then frozen until December 1947. It was agreed further that certain skilled workmen should be paid a flat rate of 1/0d a shift (to be known as the 'skilled shilling'). The miners' wage structure still reflected the terms of the 1944 agreement when the industry was nationalised in 1946. The agreement was recognised as a stopgap measure, however, and in no sense a substitute for a systematic revision of the payment system.

But, although a general trend towards national pay determination and, with single ownership, a more uniform wage structure, was perhaps inevitable under nationalisation, the problems created by the pay structure did not lend themselves to an immediate solution, and it would have been foolish, given the history of local wage determination and its formal recognition in the industry's bargaining procedures, to expect that the task of coordinating wages policy would be an easy one.

PART ONE

THE EVOLUTION OF A NATIONAL WAGE-RATE STRUCTURE

2. A NATIONAL DAYWAGE STRUCTURE

By the end of the Second World War the miners had obtained a substantial improvement in their wage position. In seven years they had moved from nineteenth to first place in the industrial earnings league.[1] Equally important, their earnings were at last underpinned by a national minimum wage. But their wage aspirations were still unsatisfied in one major respect – there had been little attempt as yet to rationalise the industry's wage structure. Uniform principles had been applied to increases during the War; but by merely superimposing a succession of nationally agreed wage elements on the existing wage levels, the wartime agreements left the original structure's character unaltered. Thus, wages still varied from coalfield to coalfield and from pit to pit, in a manner which, as we shall see later, bore little relation to the work performed or the physical efforts of the men.

One of the most notable consequences of collective bargaining in the nationalised coalmining industry has been the establishment of a standard wage scale for industrial colliery workers. This was not achieved overnight, however, but was the product of protracted negotiations. In this chapter we first consider the early difficulties encountered by the NCB and NUM in wage policy formulation. We then trace the course of collective bargaining which in 1955 culminated in the first major revision of the coalmining industry's wage structure, and outline the main features of that revision which principally concerned daywagemen.

2.1 Attitudes to wage structure reform at nationalisation

The NCB and NUM agreed, by the terms of the 'Carry-Over Agreement',[2] to honour all agreements appertaining to miners' wages at Vesting Date, thus nationalisation had no immediate impact on the industry's wages structure. Nevertheless, both sides were dissatisfied with their inheritance and appreciated that one of the main justifications of the structure's disparate form had disappeared with the colliery companies. Accordingly, they also agreed[3] that 'On and after the primary Vesting Date' they should be entitled to 'enter into an agreement . . . whereunder a new wage structure shall be established in the coalmining industry'. Despite these good intentions, however, they soon appreciated that the 'task [of revising the structure] was even more complicated than had originally appeared' and that they 'should not try to deal with the problem in all its aspects'.[4]

Neither the Board nor the Union had clear ideas on the subject of the wage structure reform. There had been talk in the NUM of a 'national wages structure', but what this meant was rather vague. The 1946 Annual Conference instructed the Union's National Executive Committee to produce a memorandum on the subject.[5] The fact that they were still asking for that memorandum in 1948[6] is probably indicative of the problems encountered. Certainly the NUM was torn between principles and practical propositions. In a speech to Conference in 1947,[7] for example, Arthur Horner[8] explained 'Our aim and our desire is to constitute on a national plane the same rate for the job from Lands End to John O'Groats: but we are not living in an unreal world, we are not facing this thing in a vacuum: we are confronting it with the full knowledge of the wide variations which exist at the present time . . . The Executive has given consideration to what we mean by a 'national wages structure'. Right away there appears a most difficult question which is – whether or not it is possible, at this stage, to have within a national wages structure uniformity as between division and division and district and district.'

It is difficult to judge NCB views on the wage structure at nationalisation. In correspondence with the writer, one observer at that time described the situation as follows: 'You will recall Shinwell's utterance about arriving at the Ministry of Fuel and Power with no "blue prints" for nationalisation after 40 years talk about it. Substitute "wage structures" and you have the situation. Certainly the National Board itself had no specific ideas about what a "wages structure" meant. Individual Board members *may* have had "ideas" – vague general ones, and if they had been expressed, no doubt they would have been a real conflicting hotchpotch.'

One factor which neither side could ignore was the cost involved in any revision. It was unlikely that a move towards wage uniformity could be achieved at the expense of the higher-paid working groups and areas.[9] This made an immediate move to national uniformity extremely expensive, given the wide variations in pay at that time. On this account alone, therefore, there was much to be said for gradualism.

There was also a problem of reconciling a more uniform wage structure with the need for more coal. The historical basis of wage determination in coalmining had allowed financial considerations to play an important role in shaping the regional wage structure.

Although the succession of flat-rate war additions to wages had produced a fairly significant contraction in that structure, Figure 2.1 shows that, as late as 1950, it still retained a pattern which reflected earlier economic considerations. That is not to say that productivity and profitability were the only factors of importance in the coalfield earnings league. The Kent coalfield, for example, had been obliged to pay higher wages than could be justified in purely relative financial terms because of the higher relative earnings in the South East generally, as well as the relatively high earnings being paid in the

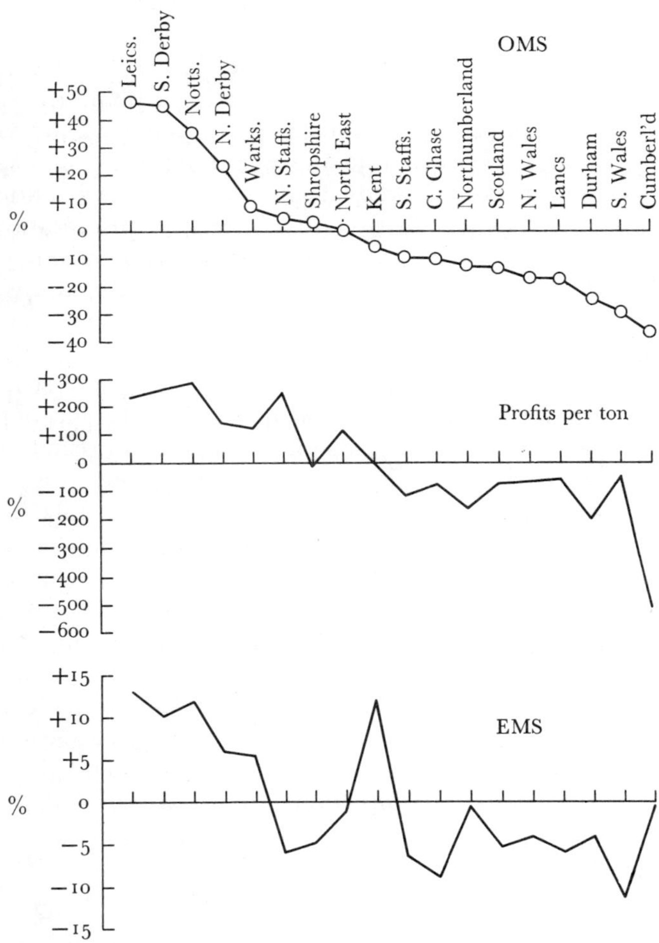

Figure 2.1 Percentage deviations of OMS, profits per ton, and EMS from the weighted national average in 1950
Source: NCB, *Quarterly Statistical Statement and Accounts.*

Midlands and Yorkshire coalfields where many of the Kent miners had their origins. But, with this and one or two other obvious exceptions, the implications of a wage equalisation policy were fairly obvious – the NCB would have to penalise further its most productive and profitable areas, on which it was to become increasingly dependent and where it could least afford disruption, in favour of its least productive and profitable. In 1948, for example, the NCB noted that the contraction of the regional wage structure during the previous decade had taken place against a widening of productivity differences between the coalfields and in the light of this was considering just how much further the process of differential wage contraction should go.[10]

The fear that too drastic an upheaval of the wage structure would conflict with its output objectives was particularly acute in the case of pieceworkers. The Board was under considerable pressure from the government to place a revision of the pieceworker's wage structure at the forefront of any subsequent wage reform and the NUM was committed to abolish piecework. As well as being aware of the threat to production which such a revision might pose, it also had to reconcile the demand for wage equality with a system which sought to relate wages to output.

The financial arrangements of the industry, which permitted cross subsidisation of pits, removed some of the economic constraints to levelling up the wage structure which had existed hitherto. Even so, unless productivity in the high-cost areas could compensate for the faster rising wage costs, and coal demand maintained a steadily upward trend, a policy which would raise wage costs fastest in the least productive coalfields had uncomfortable implications, when long term investment decisions were to be guided by considerations of relative profitability.

A pricing system based on the average cost of production in the industry as a whole, rather than the average or marginal cost of production in a particular pit or coalfield, would again protect the high-cost coalfields from some of the adverse consequences which a faster-than-average increase in wage costs might have otherwise implied for prices. But it would also tend to depress their relative levels of profitability and further increase their vulnerability in the longer term.

Thus even within the framework of the nationalised industry, one can understand some trade union leaders being hesitant to force a wages policy which could easily cause the elimination of marginal pits and coalfields in the longer run, and others, for whom these implications had little relevance, being reluctant to accept it. All these considerations favoured a gradual approach to wage structure revision.

These problems apart, wage reform had to stand alongside other important issues at nationalisation, and was eventually forced to take a back seat. Of these problems the union's long standing claim for a standard five-day working week without loss of pay was particularly pressing.[11] An agreement on this was reached in May 1947, whereby payment for what in effect would be a shift no longer worked was to be conditional upon a man having completed his full stint of five shifts in any one week. This introduced what became known as the five-day-week-attendance bonus on existing wage levels. The bonus for pieceworkers was to be 16% of normal pay and for daywagemen 20%.[12]

Another problem at nationalisation, to which the NUM gave utmost priority, was that of the 'lower-paid workers'. In 1946, for example, the NUM explained to the Prime Minister that, while they wanted a new wage structure by 1 January 1948, this should not be achieved at the cost of holding up negotiations on a new minimum wage.[13] The problem of 'the lower paid' dominated wages policy for the next eight years and became intimately linked to the problem of wage structure reform.

The lower-paid workers in coalmining were mainly daywagemen employed at the pit surface and underground. In 1947, A. Horner[14] explained that the 'main difficulties with our daywagemen ... is not so much the absolute wage as the relation of that wage to pieceworkers' wages in these days as compared with that obtained some time ago'. Table 2.1 attempts to give the problem a statistical context.

Between 1942 and 1947, approximately 60% of the daywagemen employed underground were paid no more than the minimum wage rate given in column 2. Although the underground minimum underestimates the average earnings per manshift of underground daywagemen and ignores the possibility of a 'wage drift', it is nevertheless an accurate estimate of the wages of the 'lowest paid'.

If we compare the movement of faceworkers' EMS, taken as a proxy for pieceworkers' EMS, with that of the underground minimum, it is obvious that faceworkers gained far larger wage increases than daywagemen on the minimum during the period depicted; and particularly so after 1944 – the year of 'Revision of the Wages Structure Agreement'.

We described in Chapter 1 how daywagemen gained relatively more from the 1943 minimum wage award than pieceworkers. In Table 2.1, for example, the differential in column 5 narrowed from 125.4% to 111.6% between 1943 and 1944. In addition to consolid-

Table 2.1. *Faceworker/daywage differentials 1942–47*

	(1) EMS coal face (excluding allowances in kind)	(2) Underground minimum per shift	(3) Difference between (1) and (2)	(4) Percentage Differential $\frac{(3)}{(2)} \times 100$	(5) Index of column 4 (1942 = 100)
1942	21/1	13/10	7/3	52.4	100
1943	22/11	13/10	9/1	65.7	125.4
1944	26/5	16/8	9/9	58.5	111.6
1945	29/-	16/8	12/4	74.0	141.2
1946	30/5	16/8	13/9	82.5	157.4
1947	35/5	19/2[a]	16/3	84.5	161.8

[a]Before the wage increase of December 1947.
Source: Ministry of Fuel and Power, *Statistical Digest*. Ministry of Labour, *Time Rates of Wages and Hours of Work*.

ating the wages of daywagemen and pieceworkers, the 1944 agreement sought to restore the differential between them. It also froze wage rates for a four-year period with the proviso that piece rates could be revised if a substantial change in their working methods or conditions of work warranted such a revision. The most popular explanation of the widening of the pieceworker/daywage differential between 1944 and 1945 has been that a liberal interpretation was given to applications for piecerate revisions.[15] However, our evidence suggests that much of it was caused by the 1944 agreement over-compensating pieceworkers for the erosion of their differential after 1943. This resulted from the agreement inadvertently permitting the war-time flat rates to be included in the newly consolidated piece rates. Between 1945 and 1946, the drift in pieceworkers' EMS was less than in the previous year. While the drift was increasing again in 1947, the increase in pieceworkers' EMS in that year was primarily due to the revision of pieceworkers' tasks necessitated by the Five-Day-Week Agreement. By shortening the working week for daywagemen, however, the Five-Day-Week Agreement also produced an increase in their effective shift rate. This compensated somewhat for the gain in pieceworkers' earnings. Nevertheless, as a result of the 1944 agreement, the pieceworkers' wage advantage over the lowest-paid daywagemen rose by 26 percentage points (column 4) between 1944 and 1947.[16]

The 1944 Agreement should have lasted until 1948, but, following the establishment of the Pit Conciliation Scheme, the Agreement was reconsidered in 1947. The Pit Conciliation Scheme[17] gave pieceworkers a formal channel through which to negotiate locally. An embargo was placed on pit and district wage negotiations in the case of daywagemen. The NUM was left to improve the relative wages of the lower-paid daywagemen nationally. We shall examine later the extent to which formal provision for local bargaining for pieceworkers hampered the NUM's attempt to improve the relative pay position of the lowest paid during the early years of nationalisation.

2.2 Early wage structure developments

There were seven national wage awards between 1947 and 1955, each justified, in part, by the desire to improve the wages of the lower-paid. With the exception of the award in 1952 pieceworkers gained nothing from these agreements. The agreements followed a similar pattern; in each case the national minimum wage was raised and general increases limited at the upper end by agreed ceilings. Thus, whereas the lowest-paid daywagemen received the full wage increase, workers whose wage rates were near, but below, the agreed ceilings, received only a proportion of that amount. Those workers on wage rates above the ceilings received nothing. The ceilings were raised periodically. But the general effect of this policy was to squeeze the differentials between the highest- and lowest-paid daywagemen. This policy represented a gradual move towards wage uniformity. But it also created many problems. First, it applied only to daywagemen. Second, the policy was indiscriminate of occupational differentials. Third, because daywagemen had never been defined, the policy was, with the exception of craftsmen, not applied systematically.

By 1948 the Board and union had been able to agree on a list of jobs to be designated 'skilled crafts'.[18] These were divided into two groups. The first group consisted of two grades: fully skilled men were put in grade I and were to receive a slightly higher shift rate than the remainder in grade II. In the second group there was only one grade, the shift rate being the same as the grade II rate of class I. In all cases a premium of 2s 6d a shift was payable to workers employed underground. The wages of workers whose rates were below the national wage rates of 1948 were brought up to the new rates. Others above the new rate received no increase but did not have their rate reduced.

But why did an attempt to revise the wage structure start with craftsmen? Craftsmen were a relatively small section of the labour force so the cost of revising their structure was relatively small. Even so, the rates agreed in 1948 were national minima only: the agreement did not establish wage uniformity for craftsmen on wage rates above the minima. This was to be achieved gradually through the ceilings policy as previously outlined.

The Board and union had little detailed factual knowledge of the wage structure at the time and this made the unsystematic approach to the problem during the early years of nationalisation inevitable. Committees had been established to provide more information, but, in the face of other pressures, they made little or no progress. It eventually took a hard line from the industry's National Reference Tribunal to get things underway. The Tribunal considered a revision of the wage structure of such priority that towards the end of 1952 it refused to consider a union wage claim in case it jeopardised progress on the revision of the wage structure.[19] This decision was hardly popular in the coalfields and pressure was put on the Board and union to devise a blueprint. An agreement on the first major revision was eventually signed in April 1955.

2.3 The daywage structure

It would be difficult to exaggerate the amount of work involved in the daywage structure revision completed in 1955. The first task of the committee appointed for that purpose had been to determine the nature of all the separate jobs in the industry and to draw up a list of standard job descriptions. The chaotic state of affairs was typified by the fact that, although there were only some 80–100 jobs which could exist within an average size colliery, apparently over 6,000 differently named jobs were being performed in the industry.[20] By standardising the nomenclature of these jobs so as to eliminate the many different names for the same task and then grouping similar jobs together, this list was eventually whittled down to 400. The jobs were then grouped into five underground, four surface[21] and three craftsmen grades, the work content of the different jobs in each grade being, as far as possible, of similar skills.[22] A brief description was attached to each job and this formed the basis of the so-called 'Black Book'.

The trickiest part of the whole operation was to assign a national wage rate to each grade. The survey of tasks and rates revealed that there was little, if any, consistency in the wage rates paid for similar tasks, and very often the rates for jobs in the lowest and highest grades overlapped quite considerably. Inevitably many local and traditional wage differentials had to be disturbed. To minimise the unrest which a disturbance of this nature might induce, it was agreed that no man would have his wage rate reduced.

Four important points must be appreciated about the wage rates which were eventually agreed upon and assigned to each grade.[23] First, they applied only to daywagemen; pieceworkers were graded in 1955, but their wage rates were left to be determined at local and district level. Second, the new wage rates were to be standard throughout the industry. Third, they were to be all-inclusive and 'not ... subject to any additions in the way of bonuses, flat rates or allowances whether national or local, temporary or permanent, except for temporary allowances in respect of men working in wet conditions, the five-day-week bonus, and rent allowances where the rent allowance is made as an addition to wages'. Fourth, they were 'not to be altered in any way except by national agreement'.[24].

In estimating existing wage rates payable to adult daywage workers, the Board and union consolidated the numerous wage items from which the wages of adult daywage workers were compounded, up to the time of the agreement, irrespective of their name or purpose (bonus, allowance, 'skilled shilling', 'porter make-up', etc.), and whatever form they took (flat rate, allowance per shift, allowance per week, bonus per shift, bonus per week, etc.). Only temporary wet payments, the five-day-week bonus and rent allowances were excluded from this calculation.

When all the day rates had been consolidated in this manner they were then increased by 1s per shift, subject to ceilings over which no increases would be awarded. Craftsmen were excluded from the process, however, because in their case the standard grade rates already set the limit to their wage rate increases. Standard grade rates for daywagemen, other than craftsmen, varied from 27s 9d to 31s 9d underground and the ceilings from 30s 0d to 39s 6d. For surface workers the grade rates varied between 24s 5d and 27s 5d and the wage ceilings from 27s 6d to 34s 0d. The wage structure thus provided for very little wage differentiation at either place of work.

In cases where, after applying the above procedure, a worker's shift rate was still below the appropriate new standard rate, the shift rate was further revised so as to bring that rate up to, but not above, the standard grade rate for the job grade. The same procedure was adopted in the case of those craftsmen whose rates were below the new standard grade rate.

There were, of course, many workers receiving shift rates above the standard grade rate. It was agreed that the workers concerned would retain their *existing shift rate so long as they remained at the colliery at which they were working at that time and so long as they were not downgraded.* These rates were to be known as 'personal rates'. It is important to appreciate that a 'personal rate' 'belonged' to a particular workman and, should he retire or move to another colliery, could not be claimed by the successor to his job. The only instances in which the worker would not retain his personal rate would be when he was temporarily called upon to perform a job whose standard grade rate was below his personal rate, or if, without substantive upgrading, he was called upon to perform a grade of job with a rate higher than his 'personal rate', or was downgraded to a job for which the standard rate was lower than his 'personal rate'. As men retired or changed jobs, therefore, personal rates were expected to disappear and in time all daywagemen would be on the standard rate. Table 2.2 shows that when the wage structure was implemented, some 21% of daywagemen were on wage rates which were above the standards agreed. Of these 18% were on personal rates, the size of which lay between the standards and the wage ceilings, and just over 3% were on personal rates in excess of the wage ceilings.

The high proportion of craftsmen on standard rates was due to the

Table 2.2. *Distribution of daywagemen in the revised daywage structure in April 1955*

	Percentage of group on:			
Place of work and category of worker	Standard grade rates	Above standard but on or below wage ceiling	Above the wage ceiling	Total
Coalface				
Non craftsmen	58	27	16	100
Craftsmen	86	14	0	100
Total	61	25	14	100
Elsewhere underground				
Non craftsmen	75	21	4	100
Craftsmen	91	9	0	100
Total	78	19	3	100
Surface				
Non craftsmen	83	16	1	100
Craftsmen	96	4	0	100
Total	86	14	0	100
Grand total				
Non craftsmen	76	20	4	100
Craftsmen	93	7	0	100
Non craftsmen and craftsmen	79	18	3	100

Source: Calculated by the writer from NCB Working Papers.

earlier revision of the craftsmen's wage structure. Nevertheless, the proportion of craftsmen on standard rates did vary from 85.7% at the coalface to 96.5% at the surface. At the time of the daywage revision in 1955, however, there were no craftsmen on rates in excess of the wage ceilings. Among daywagemen other than craftsmen, there was a much wider dispersion of wage rates. Although the numbers were small, only 57.6% of men at the coalface were on standard rates, while at the surface over 83% were on standard rates. The majority of the workers on wage rates above the wage ceilings were employed in the highest grade at the coalface and elsewhere underground.

The existence of personal rates was likely to cause discontent among men on the standard grade rates. Thus although personal rates would have disappeared gradually as men changed jobs, a more positive approach was required with respect to personal rates. It was intended, therefore, that wage rates in excess of the wage ceilings would be frozen while those between the standard and the wage ceiling would only be increased by an amount which would not raise them above the ceiling *fixed in 1955*.[25] If this policy had been followed, the time required to absorb personal rates into the wage structure would have depended on the size of the national pay increases and the distribution of men between personal rates.

The frequency distributions of non-craft daywagemen in 2/- wage bands in April 1955, shown in Figure 2.2 reveal that, with the exception of daywagemen in grade I underground, the largest proportion (85%) of men on personal rates in each grade were concentrated in a wage rate band of 1d–5/11d above the appropriate standard grade rate. This suggests that if the wage ceilings had been held at the levels agreed in 1955 an annual average wage rate increase of 2/- per shift[26] would have guaranteed that over 98% of daywagemen would have been on the appropriate standard wage rate within three years.

Some 93% of the industry's daywagemen received a wage increase under the terms of the 1955 daywage agreement. Given the wide wage disparities which existed before the agreement, though, it was inevitable that the proportion receiving a wage increase should differ markedly between wage districts and between men working in different parts of a pit. From Table 2.3, for example, we can see that the average proportion of men receiving a wage increase from the agreement ranged from 99%, in North Wales, North Staffs and the Forest of Dean, to 73% in North Derby. A larger proportion of surface daywagemen received increases than daywagemen employed underground (99% and 95% respectively), although the proportion receiving increases again varied between districts. Daywage rates had

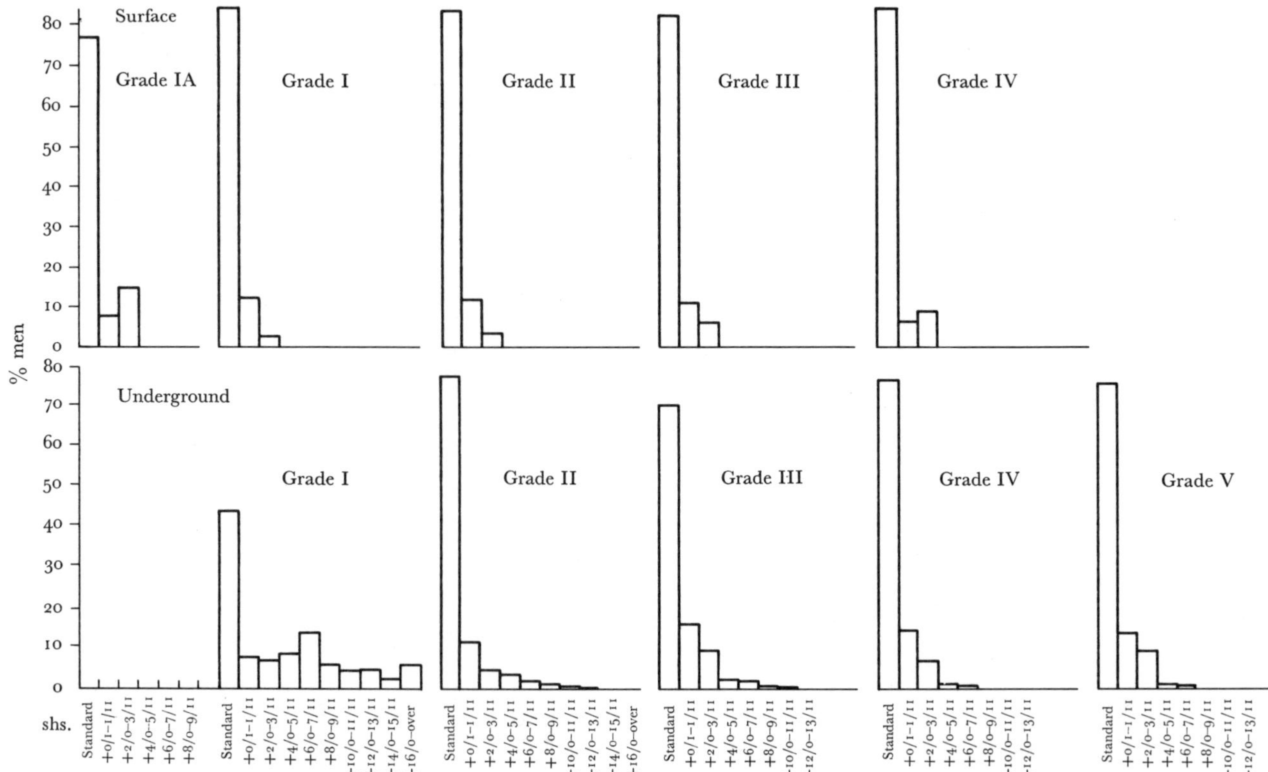

Figure 2.2 Distribution of daywagemen in the revised wage structure in April 1955
Source: Constructed by the author from data supplied by the NCB.

Table 2.3. *Percent of adult daywagemen receiving increases in consolidated EMS following revision of the wages structure agreement, April 1955*

Wages district	Coal face %	Elsewhere underground %	Surface %	Total %	Ranking of total[a]
Scotland	82	97	100	95	9
Northumberland	47	97	100	93	10
Cumberland	46	86	100	88	15
Durham	71	97	100	95	7
Yorkshire	75	97	100	96	6
Lancashire and Cheshire	69	95	100	92	11
North Wales	100	98	100	99	3
Notts	49	96	87	87	17
North Derby	16	84	99	73	20
South Derby	33	79	97	78	18
Leicester	65	87	99	89	13
Cannock Chase	49	89	100	88	16
North Staffs	95	100	100	99	1
South Staffs	27	95	99	95	8
Shropshire	53	92	97	88	14
Warwick	47	84	99	89	12
South Wales	97	98	100	98	4
Forest of Dean	98	99	99	99	2
Somerset	94	99	100	98	5
Kent	15	72	100	74	19
Great Britain	72	95	99	93	

[a]This refers to unrounded total percentages.
Source: Data supplied by the *National Coal Board*.

been most varied among the few men employed at the coalface. There 72% received increases, but in some districts fewer than 20% of face daywagemen received increases.

The percentage of men receiving increases from the 1955 daywage agreement is a reasonably accurate reflection of the relative impact of the agreement on wage levels as between wage districts. In the short run, of course, no one lost. The *average* increases in shift rates ranged between 2/1, in areas such as South Wales and North Staffs, to 6d in Nottinghamshire and Kent. In the long run the lower-paid areas would again gain most from the agreement, as they had fewer personal rate and 'mark-time' men (the name given to people on personal rates in excess of the wage ceilings).

At a special conference called on 18 March 1955,[27] the revised structure was approved, by 520 votes in favour to 204 against. The major opposition to the agreement came from the traditionally high-wage areas of the Midlands. But, in introducing the debate on the new wage structure, Arthur Horner[28] made it quite clear that although 'what we did was not the Law of the Medes and the Persians in respect of the wage structure for daywagemen, we are insistent that so far as we are able to, we are going to hold on to it, except where individual examples are shown in which variations ought to be made. But never, never can we agree, if we believe in this wages structure, that the Coal Board should concede to any pressure in any particular division to make excess payments under threats or anything else, and the Union will be as strong as even Mr Sales [the NCB Director of Industrial Relations] in trying to maintain this thing which was brought into existence after so long and laborious work.'

But opposition to the agreement was not confined to the wage issue. Prompted by a large proportion of the lodges in his area rejecting the agreement against the advice of the Area Executive, Mr Wynn of Derbyshire discovered that 'insufficient discussion in the branches'[29] was a contributory factor and that 'in some cases there was a categorical rejection of the agreement purely because they [the Lodges] had not been sufficiently informed on the matter being considered'.[30]

There were undoubtedly many detailed points in the new structure which raised considerable disagreement. Some disputes concerned the evaluation of particular jobs. However, most of these were successfully resolved through a disputes committee set up for the purpose.

The agreement's treatment of craftsmen proved more contentious. A comparison of the NUM's original grading proposals[31] with those of the NCB, for example, revealed that, whereas the NUM requested that 83.7% of craftsmen be placed in the top grade, the majority of craftsmen were eventually allocated to a lower grade. This exacerbated the feeling among craftsmen that their wage position was low, in relation both to other workers in mining and to craftsmen outside mining. This became one of the main foci of discontent in following years.

2.4 Underofficials and weekly paid industrial staff

The 1955 Agreement for daywagemen inevitably distorted their relationship with the men who supervised them and precipitated a rationalisation of the wage structure of underofficials and weekly paid industrial staff.

At nationalisation the wages of underofficials – overmen, deputies and shotfirers – varied from one district to another and frequently between pits in the same district. Deputies and shotfirers were then paid by the day, while overmen were mainly paid by the week.

The Board and the unions concerned (the NUM and NACODS) had agreed that it was important to maintain the relationship between the wages and conditions of services of their junior colliery officials and those of mineworkers. Accordingly it was agreed in 1947 that deputies' wages should be reviewed, on a divisional basis, and subsequently improved by adding not more than 5/- to shift rates, so long as they did not exceed the *average* shift earnings of pieceworkers in the division or wage district.[32]

It was agreed further that 'Overmen were not to be paid less per week than a Deputy (at the same colliery) would be paid for working six shifts, i.e. five normal shifts and one overtime shift.'[33] This, however, was only a minimum below which the earnings of overmen should not fall. Subject to the condition that they should not earn more than £100 a year more than a deputy, the actual wage rates of overmen were to be settled by the Divisional Boards in the light of local circumstances. Shotfirers' wages were unaffected, as they were generally adjusted in the light of wage increases given to mineworkers.

The task of establishing a national grade and wage structure for underofficials began in 1952, following new regulations for the qualifications of underofficials.[34] And in August 1952 it was agreed that *all* underofficials in the industry should be paid a weekly wage and not, as in some instances previously, a daily rate. The rates were expressed as ranges so that the divisions could retain some discretion to pay according to local circumstances. So, although the agreement was a move towards wage standardisation, it hardly established complete uniformity. A further move towards uniform national wage rates for underofficials followed in January 1956, when the weekly wage rates of deputies, grades I and II, and shotfirers were (with the exception of those working in the East Midlands, Warwickshire, Kent and Yorkshire, where the rates were to be slightly higher) standardized throughout the coalfields. Uniform wage *ranges* were also established for overmen, although these were a little higher in the aforementioned divisions.

The wages structure of weekly paid industrial staff (WPIS) was also reviewed in 1956. Before nationalisation, these wages were settled by direct arrangement between employer and employee rather than collective agreement. Although the wages of the WPIS were to be reviewed locally, it was agreed in 1947 that this should be done in the light of general wage changes in the coal industry.[35] When, in 1954, preparatory work began on the wage structure of WPIS, the problems were in many respects even more onerous than in the case of daywagemen. First, over 837 different job names existed in the industry, many of which covered similar jobs. This list was eventually whittled down to 65 (48 for collieries and 17 for central workshops). Second, WPIS were often treated as daywagemen. These anomalies had to be corrected without prejudice to wages. Third, there was no working definition of a foreman. This was eventually resolved by an omnibus definition in terms of responsibility. These, and other problems, delayed the establishment of a uniform wage structure for WPIS until August 1956.

Three categories of workers were distinguished in the revised structure[36] – engineering services, general services and specialist services. The first category had four grades, the second three grades and the third four. The revised day rates[37] were 'all inclusive' and standard in the same way as that specified in the daywage agreement of 1955. Similarly, it was agreed that consolidated weekly wages in excess of the standards should continue as 'personal rates'.

2.5 Summary

A lack of clear ideas on the form a revised wage structure should take, and the pressure of other events, favoured a gradual approach to wage structure revision. The Board and union did achieve a greater uniformity in the wages of craftsmen as early as 1948. But it was not until the agreement of 1955 that one could describe the national approach to the wage structure as anything other than pragmatic. And the revised daywage structure was as much a product of the pressures which the National Reference Tribunal put on the Board and union as it was of the latter's good intentions. Even then the Board and union did not go as far as the National Reference Tribunal would have wished; although every job in the industry was graded in the 1955 structure, pieceworkers' wages, which had been singled out by the National Reference Tribunal as being in particular need of reform, were relatively unaffected by the new standard shift rates.

The daywage structure adopted the NUM principle that men should be paid according to the jobs they performed rather than according to the pits in which they worked. The adoption of this principle was facilitated by the ownership pattern imposed upon the industry by nationalisation and by the guarantee that no man should suffer a wage cut for the cause. As some areas, nevertheless, stood to gain much less from the agreement than others in both the short and long run, it says much, for the NUM and its members, that the agreement was received so peaceably in the coalfields, as well as for the consideration given to the problems by the wage-structure planners.

While the general character of the revised daywage structure was determined by equity considerations, cost was a deciding factor in determining the pattern of wage rates laid down in the agreement. The revised daywage structure would have created anomalies between the members encompassed by the reform and their immediate supervisors, had the latter's wage structure not been revised. The cost of the necessary adjustment also had to be reckoned with in the overall budget constraint. While the Board recognised that the pattern of differentials was far from ideal, it could not afford larger differentials in the short run. Cost considerations also made it impossible to achieve complete wage uniformity within labour grades overnight. If the underlying principles of the agreement had been adhered to, however, it should have been possible to achieve near uniformity of wages within the space of three to four years. We shall examine subsequent developments in the daywage structure in Chapter 5.

3. THE PIECEWORKERS' WAGE STRUCTURE: THE INHERITANCE

The first major revision of the coalmining industry's wage structure standardised the work content of various occupations, grouped them into grades and allocated a national time rate to each grade. All colliery manual workers were assimilated into the new grading structure but only daywagemen were affected by the changes in wage levels which accompanied the 1955 agreement: pieceworkers – some 42% of the manual labour force in 1955 – were only affected insofar as the standard day rates became minima below which their wages, still determined on a local and district basis, could not fall, and by the extent to which they would subsequently affect waiting-time and make-up payments when their own normal work was unavailable. As locally agreed fall-back rates for pieceworkers were usually well above the national minima, one can conclude that, apart from any indirect influence which the change in the time rates may have had on the tempo of local applications for piecerate revisions, the effect of the 1955 wage agreement on pieceworkers' wages was minimal.

The piecework system which the industry inherited at nationalisation was extremely complex, which probably explains why there has been little analysis of its effects in terms of wage levels. This writer has been exceptionally fortunate in having had access to the papers of the NCB working party, established in 1956, to examine the possibility of reforming the system. Together with other information collected at a later date, the working papers provide a fascinating picture of the old piecework system. Before we consider the events which culminated in a national revision of the piecework system in the late 1960s and early 1970s, therefore, we shall examine, in this chapter, some features of the early piecerate system which, in addition to having been the source of obvious discontent among the miners, eventually led the NCB and NUM to discard their ideas for revising the system in the mid 1950s.

3.1 Piecework, price lists and contracts

The NUM, and the Miners' Federation before it, had always opposed piecework. At the 1946 Conference, for example, the national executive were instructed to secure its complete abolition.[1] But the union leaders appreciated that, so long as physical output was still dependent to some extent on a man's effort, attempts to abolish piecework would meet with little success at a time of fuel shortage. Nor could the union ignore the fact that, although piecework inevitably led to earnings differentiation, this might be justifiable if men worked unequally hard. Thus early discussions on a revised piecework system were aimed at its reform rather than its abolition. The union's main objection to the piecework system was its complexity, inequality and lack of amenability to central direction – an objection shared by the NCB. Let us consider the substance of these arguments in more depth.

In theory the scope for applying piece rates in coalmining was restricted, because very few jobs entailed work which could be easily measured. In the absence of any central directive it would have been surprising, though, if local managements were consistent in their choice of jobs to which piece rates were applicable. And the fact that small pockets of pieceworkers were found in every task listed in the 1955 grading structure probably reflects the misuse of piece rates rather than their general suitability. Of the pieceworkers employed at that time, 89% were found to be concentrated in only twelve jobs. And of those twelve jobs only two – tunnelling and back ripping – were non-coalface operations. Thus, whereas 83% of the workers employed at the coalface were on piece rates, only 0.7% of the workers employed at the pit surface, and 19% of the workers employed 'elsewhere underground' (i.e. between the pit bottom and the coalface), were on wage incentives in 1955. There is little danger, therefore, of our falsifying a description of piecework as it has operated if we restrict our discussion to coalface operations.[2]

In the mid 1950s the hand-got system of mining was practically obsolete and most deep-mined coal was cut by machines. But power loading and its supporting system had made little progress; the actual *loading* of coal (filling) was still done by hand, as were most of the other ancillary operations at the coalface. The changeover from hand-got systems to machine mining, however, necessitated a dramatic change in work organisation, in that all the operations – ripping, packing, filling, support setting, etc. – which had previously been done by one man (the collier) became specialist tasks which had to be performed in an agreed sequence over the three shifts. The number of men required for each operation varied with the length of the coalface, but the deployment on a 110 yard handfilled (machine-cut or conventional) face shown in Table 3.1 was not untypical.

Table 3.1. *Tasks and manpower allocation on a 110 yard conventional longwall coalface*

First shift	Second shift	Third shift
boring (1 man) coalcutting (2 men) ripping (4 men)	stripping and filling (10 men)	packing and withdrawing supports and conveyor moving (5 men)

Under the hand-got system of working, the collier would have been paid a comprehensive rate based on the tons of coal filled, but usually with allowance for such ancillary operations as were 'necessary to perform his composite task'. Where the piece rate did not take the other work into account, the collier's pay was supplemented by a system of allowance payments, which over time tended to harden into 'custom-and-practice'. For a machine-cut face, however, most collieries had a piece rate for each of the operations listed in the table. With the exception of filling, most of these rates would be 'all-in', in the sense that the rate would be fixed for the job and then payment would be made to the group according to the measured quantity of work performed in (say) a week. Fillers might be on a variety of contracts, but mostly they were on individual contracts or contracts covering small sections – i.e. small groups of adjacent fillers on the same face. Each colliery might, of course, have anything up to ten faces operating at the same time, in different seams. Because the main factors affecting a workman's efficiency – the thickness of the seam, the friability of the coal, the seam gradient, the amount of clod (dirt), temperature, water and the conditions of the roof and floor – might vary quite substantially, a separate price list was often drawn up for each face or seam.

In coalmining a variety of piece rates were feasible because of the different ways in which work could be organised. Thus, face operations performed by pieceworkers in adjacent working positions were often lumped together into a single contract which fixed a piece rate for a group of operations, but usually expressed the price in terms of one operation only, rather than setting out the individual piece rates for each operation included in it. In appearance, then, the contract was for a single operation, but in fact it covered a variety of operations and applied not to an individual but to a group of men. Even with half a dozen or so tasks, the number of combinations of operations was large, so that a whole range of contracts was possible.

One quite common grouping of operations occurred when the work of withdrawing supports was brought into a conveyor-shifting or packing contract. In this instance the men were covered by a *combined or composite contract.* A large number of pieceworkers all doing both kinds of jobs then shared the money obtained from the contract. Sometimes, though, *all* the facework on a single shift was brought into one *all-in* contract. In this case a single index of results – usually the output of coal per manshift – was taken to calculate face earnings. The results from individual operations in this contract would not be measured separately. More rarely, the contract covered all the work done on all three shifts, on the same face. Like the former contract, this was supposed to encourage team spirit and induce cooperation between the workmen on different shifts. Usually the total face earnings were divided equally among the men working it, but sometimes the men and the chargemen would determine the rate to be paid each man, or each might receive the district or normal rate for the job and a share of the remaining pool of earnings based on the time worked on the face during the week. At the other extreme from the 'all-in' contract, a contract was often made for the individual worker doing just the one type of work, which might be only a fraction of that necessary on the shift to keep the cycle going. Getters and fillers on machine-cut faces were usually parties to *individual contracts* by which, having determined the rate for the job, payment was made to the individual on the basis of task performance. This type of contract could be complex if one man was required to do more than one operation. Sometimes small groups of men (say 5–6) doing the same operation, or a mixture of two operations, would be brought together in a *sectionalised contract.* By keeping the group small the difficulty which emerged in the larger 'composite' or 'all-in' contract, of seeing a connection between pay and personal exertion, was minimised. Since the men themselves determined the composition of the team, this contract tended to get men of similar temperaments and capacities working together. These contracts were popular with the men. Local NUM officials, however, were sceptical of them because, in the process of selection, the older men of less physical ability tended to get separated out and subsequently received lower earnings than younger workers. They also felt that managements tended to take the tasks achieved by the best team as the standard when negotiating the contract.

The medley of contracts may be more clearly visualised with the help of Table 3.2. They could be large or small, according to the number of men in them and simple or complex, according to both the number of men and the variety of operations covered.

As the industry moved from the hand-got stall system of mining, the popularity of sectionalised and 'all-in' contracts increased at the expense of individual contracts, as management attempted to combine

Table 3.2. *Piecework contracts*

Number of operations covered by the contract	Types of contract — Number of men in the contract: *One*	*More than one*
One	Simple individual contract	Simple group sectionalised contract
More than one	Complex individual contract	Complex group contract: composite and 'all-in' contract

the advantages of incentive and cooperation. In 1957, 52% of all coalfaces were worked under sectionalised contracts and 23% under 'all-in' contracts.[3] Output per manshift (OMS) tended to be highest on 'all-in' contracts, but one cannot attach too much significance to this, as they tended to be most popular in the East Midlands where the conditions under which coal is mined are also the most favourable. As a generalisation, however, output was higher the larger the group covered by a contract. Because some contracts attempted to equalise wages between men doing similar and different work while others did not, one man's earnings in relation to those of another, doing similar or dissimilar work, would inevitably depend on the type of contract under which he was working.

Price lists have traditionally been negotiated at pit level and with due regard to the conditions under which men work, which alone would account for some of the diversity of piece rates between men, collieries, seams and faces. Even a price list tailored to local circumstances, however, could not possibly cope with the variety of conditions and additional tasks with which men had to contend from day to day. These variations were catered for instead by an elaborate allowance system.[4]

Allowance payments fell into three broad categories. First, there were allowances to cover variations in seam conditions – such as water, heat, dust and variations in the seam, such as rolls, faults and washouts – which would adversely affect a worker's performance. Next, there were allowances for work performed in addition to that stipulated in the price list. Both these types of allowance were negotiated locally, between the worker and his immediate supervisor (the deputy or overman), and, as one can imagine, they were the subject of much haggling and dispute. It is almost impossible to grade the severity of many of the 'adverse' conditions for which they were paid: thus the amount agreed was invariably a 'hit-and-miss' affair.

Nor, because adverse conditions are always judged as relative to the normal conditions obtaining in any one pit, was the same allowance likely to be paid for similarly adverse conditions between pits. In the case of both types of allowance, it was almost impossible to check whether they were paid for genuine reasons or not. The same problems applied to the third type of allowance payments, namely those for deficiencies in organisation, such as a failure to maintain satisfactory roof control, supplies to the coalface, sequence of face operations, day to day deployment of men and an efficient transfer of men between various tasks in the case of sickness, accidents and voluntary absence, etc. These can be differentiated from the allowance payments mentioned above because they were payable under national agreements. Again, though, it was almost impossible to check whether these were applied consistently.

Table 3.3 cites an example of how allowance payments served to complicate a pieceworker's wage packet, in the case of a Scottish filler.

In Yorkshire, there were over 40 recognised allowance payments which could be given over and above items provided for in the basic contract in 1961. The precise number of allowances obviously depended on the detail of the contract. Table 3.4 shows the percentage of earnings drawn from the basic contract in 1961 on the faces operating at that time in Yorkshire.

Table 3.3. *Wage composition of a Scottish filler in 1965*

Wage item	EMS s	d
A. *Basic contract, per shift*	37	9.37
B. *Allowances*		
Roof allowance	1	6.42
Cuts in clay	3	3.66
Height		0.59
Solid core		2.47
Casting		1.52
Core shifts		5.11
Other work	1	6.73
Channel		1.83
Broken roof		9.88
Waiting time (payable by national agreement)		10.23
Long throw		0.32
	46	10.05

Source: NCB Scotland

Table 3.4. *Payments from tonnage on Yorkshire coalfaces 1961*

Percentage of EMS from basic contract	% of faces
less than 50%	16
50–59%	24
60–69%	30
70–79%	17
80% and over	13

Source: NCB Yorkshire

Thus on 16% of the faces, the workmen earned more from allowances than they did from the basic contract, and on only 13% of the faces were allowance payments less than 20% of earnings.

3.2 Task performance and wage rates

While the foregoing discussion should leave the reader in no doubt as to the complexity of the piecerate system inherited at nationalisation, it is less easy to prove that the resulting wage levels were inconsistent. Under a piecerate system one does not, of course, expect wage levels, even for the same job, to be uniform between one worker and another; wage levels will vary if only because effort inputs vary between workers. But trade unionists tend to argue that equal work should receive equal pay. And this was certainly the criterion according to which the miners' leaders thought their wage structure to be inconsistent. The reports of the working party established to investigate the operation of the piecerate system in 1957 do facilitate a somewhat limited appreciation of this and related problems.

Figure 3.1 gives some idea of the extent of wage (EMS) differences of colliers, rippers and coal cuttermen. The distribution of EMS is fairly symmetrical in all three cases. But the scatter of earnings is also large, particularly in the case of rippers and coal cuttermen. Thus, whereas colliers' earnings ranged from 32/- to 80/- a shift, those of rippers ranged from 34/- to 99/- per shift, and those of coal cuttermen from 34/- to over 100/- per shift. Again, whereas over 50% of faces gave colliers an average EMS of between 50/- and 60/- per shift, only 37% of ripping lips and 34.5% of faces gave the same range of earnings for rippers and coal cuttermen respectively.

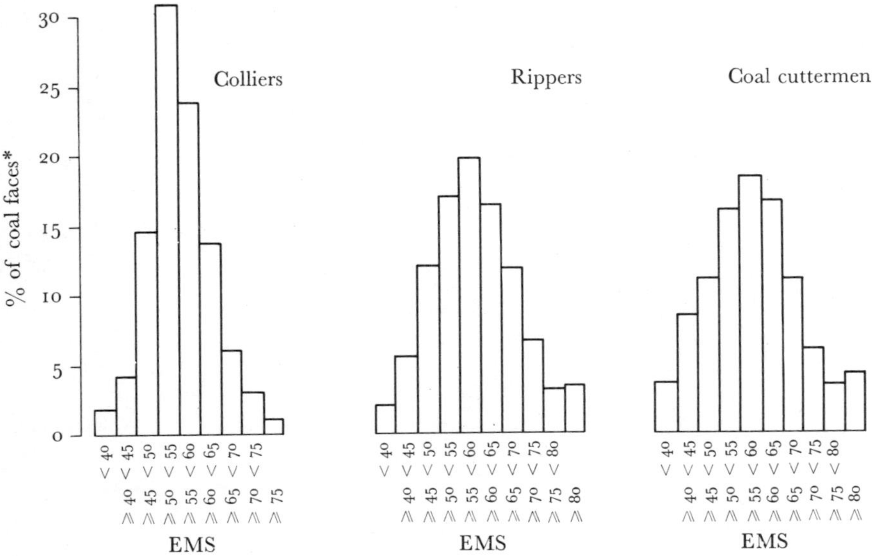

Figure 3.1 Frequency distribution of average earnings per manshift for colliers, rippers and coal cuttermen in 1956

*Ripping lips in the case of rippers.
Source: NCB, Working Party Reports.

Two questions follow from our description so far. First, did variations in piecework earnings reflect variations in worker efficiency and second, were labour evaluations consistent as between jobs?

The first question can be answered in part with the help of Table 3.5, which shows the relationship between colliers' EMS and task performance measured in terms of cubic feet of coal handfilled on to conveyors per manhour.

Superficially, the data indicate a very strong relationship between earnings levels on the one hand, and task levels on the other. But this requires some qualification. First, it is immediately apparent that the range of EMS was proportionately far smaller than the range of tasks achieved. Thus, whereas a task of 25 cubic feet per manhour tended to attract, on average, payment of 50/- per shift, one of 50 cubic feet was rewarded with only an additional 3/3d per shift. Figure 3.1 shows that over 50% of faces were paying colliers between 50/- and 59/11d per shift. We can now see that this range of payment was for a task output which varied between 25 and 85 cubic feet per manhour.

The above characteristics of the relationship between mean task

Table 3.5. *The relationship between colliers' tasks and earnings per manshift in 1956*

Representative task per manshift[a] Cubic feet	Mean representative earnings per manhour[a] shillings
Under 25	35.8
25 and under 30	50.1
30 and under 35	51.1
35 and under 40	50.9
40 and under 45	51.7
45 and under 50	52.6
50 and under 55	53.2
55 and under 60	54.0
60 and under 65	55.2
65 and under 70	55.9
70 and under 75	57.1
75 and under 80	57.6
80 and under 85	59.6
85 and under 90	61.4
90 and under 95	61.5
95 and under 100	62.0
100 and under 105	64.0
105 and under 110	65.0
110 and over	64.0

[a]The figures for tasks and earnings refer to a normal or 'representative' working week. In fact there was a very close correspondence between representative and actual statistics.
Source: NCB, Working Party Report, 1957.

performance and EMS can be expressed more generally in the following regression equation (where E represents mean earnings per manshift in shillings and T the representative task per manhour in terms of cubic feet of coal).[5]

$$\log_e E = 2.950 + 0.2591 \log_e T$$
$$(0.115) \quad (0.0277)$$
$$R^2 = 0.838 \quad n = 19$$

This tells us that a 10% increase in task performance was associated, on average, with an increase of only 2.6% in average earnings per manshift. It would appear therefore that, although on average colliers' earnings were broadly related to output performance, there were pressures to pay similar wages for widely differing task performances. Again, earnings tended to be very sticky around the 50/- per shift mark. At that time the minimum shift rate for

Grade I face pieceworkers was 34/1d. This suggests that an informally agreed wage norm was operating in most coalfields to keep EMS well above the national minimum. We shall return to this shortly.

But a general relationship between output and earnings levels cannot be taken as proof of wage structure consistency: the relationship might simply reflect the employer's historical relative ability to pay. The test of consistency in a piecerate payment system must be whether differences in workers' *physical inputs*, and/or skill, are rewarded in a systematic fashion. Obviously, this can only be tested by relating pay levels to task norms established according to a common set of principles.

The derivation of a national schedule of task norms was no mean feat and occupied the NCB's working party on the pieceworkers' wage structure for the best part of two years. Very briefly the committee had to analyse all the variables which had an effect on performance on the 4,700 coalfaces then operating and apply derived coefficients to the task norm under standard conditions. With this information, actual task performance could be compared with nationally derived task norms and related subsequently to the payment actually made for that task, to give an accurate picture of the existing wage structure. It is not possible to reproduce here all the findings of the NCB's working party, and the results obtained. Nevertheless, its analysis of colliers' wage levels and tasks, conducted along the above lines, does provide extremely useful illustrative material in the context of our discussion. Some results of this analysis are summarised in Figure 3.2.

The term 'task' in Figure 3.2 refers to the actual task achieved by colliers on coalfaces expressed as a percentage of the nationally computed standard task norms. A value of 1.0, for example, means that actual performance equalled that of the nationally computed norm, whereas a value of 0.5 indicates that actual performance was only half the national task norm and a value 1.5 a coalface whose performance was 50% above the national norm.

The survey showed that in the case of colliers the distribution of task performance was fairly symmetrical – about half the faces achieved more than the norm and about half less. The range of performance, however, was quite large: from under 35% to more than 200% of the norm. But if the 5% of faces at either extreme were ignored as exceptional, performance ranged from about 60% of the norm to 140%, 90% of the faces falling within the broad band of norm ±40%. In fact, over 70% of the faces achieved a performance within ±25% of the task norm, 40% were within ±10% of the

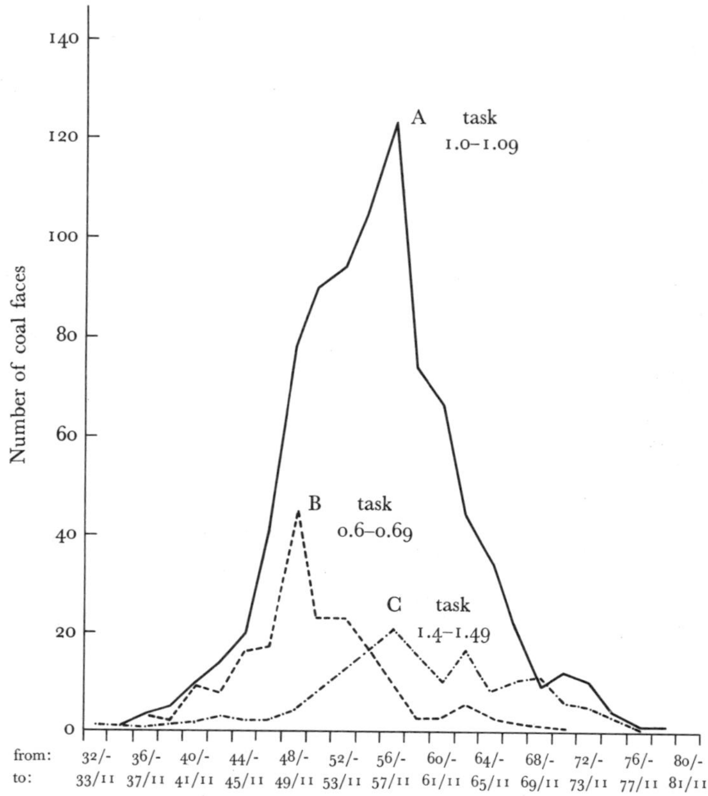

Figure 3.2 The relationship between colliers' tasks and earnings per manshift in 1956
Source: Derived from NCB, Working Party Report on Colliers' Wage Levels.

norm, and just under 20% achieved a performance which was practically equal to the norm.

Given that variations in face conditions were allowed for when computing national task standards, one would expect a consistent piecerate structure to yield a common level of EMS at any given level of task performance, as measured above. Furthermore, one would expect to find a consistent proportional relationship between different task and EMS levels. When colliers' shift rates were brought into the picture, the inconsistency of the system was displayed by the deviation from a common standard, thus proving the absence of any strong degree of coordination over norms of money and task: whether the range of EMS for a given task performance or the range of performance for a given EMS was examined, the variation in the structure and the comparative absence of payment related to performance became apparent.

In Figure 3.2, the lines A, B, C represent, in turn, common task performances of 1.0–1.09, 0.6–0.69, and 1.4–1.49 respectively. In the case of A, a task performance equal to, or slightly above, the national norm, was paid an EMS which ranged from 32/- on the one hand to 77/11d on the other. Similarly, a task performance of 0.6–0.69 of the norm (B) was awarded an EMS ranging from 36/- to 72/-; and a task performance of 1.40–1.49 of the norm (C) was paid within a range of 32/- to 76/- a shift. Alternatively, if one chose a particular value of EMS, say 54/- to 55/11d, this payment, like many others, was made for the complete range of task performances depicted in our diagram (and over a far greater range if we cared to plot more extreme task performances).

Nevertheless, as inconsistent as the wage structure was in detail, it is interesting to note that, as in Table 3.5, the modal value of EMS is higher with improved task performance. And this relationship was maintained, with minor deviations, throughout the various task performances examined. Again, though, the differences in the modal values of EMS are rather small when compared with differences in face performance. To explain this and many of the features of the system, one needs to examine attitudes to wage setting before and after nationalisation.

The private colliery owners were very sensitive to the market for their product when negotiating wages. The market for coal differed widely between regions – and even within very small areas, which explains the owners' fight against uniformity in wages. The miners had their ideas of a 'fair day's wage' but they too were reasonably realistic, and this notion would be related to what the market could bear at any time. It was, therefore, a variable notion both between pits at a moment in time and over time. If the market changed, a variation in the price list or allowance payments would be sought either by the men or the colliery owners, depending on which way the market was drifting. In these circumstances wages were negotiable only within well-defined economic limits. To force the wage above a viable level would have led to unemployment in the short run, and in the long run also, if the pit was forced to close through uncompetitive price bidding. Given that pit profitability was, and still is, highly dependent on productivity, one might, in these circumstances, expect two things. First, the owners would want to retain an incentive element in pay. Second, the wage-price per ton would vary between pits with their ability to pay. Although both these factors would give rise to considerable earnings differentiation between pits and between coalfields, some sort of positive relationship between earnings and productivity would be maintained in both instances, as the tonnage cost was the object of wages policy.

With nationalisation the attitudes to piecerate fixing changed dramatically. The Coal Industry Nationalisation Act[6] charged the NCB with the duty of 'working supplies of coal available at such quantities and size, and such quantities and at such prices, as may seem to be the best calculated to further the public interest in all respects, including the avoidance of any undue or unreasonable preferences or advantage' and to secure that the revenues of the Board shall not be less than sufficient for meeting all the outgoings properly chargeable to revenue account in an average of good and bad years'. This made it no longer necessary to produce all the coal at a profit. The consumer no longer had to pay the true marginal cost of producing coal as the price was now to reflect the average cost of the good and bad seams combined. If this policy could be applied to coal prices it seemed logical to local management and union officials that it should also be applied to wage determination.

After nationalisation there was a gradual adoption of earnings norms in local bargaining. This development was purely a local initiative; at no time during the early years of nationalisation was the policy given any official blessing from national headquarters, although by the late 1950s it was a common feature of district negotiations. Consequently, no guiding standards were given either. To some extent, the standards were provided by the ruling piece rates, but their use as guides soon became eroded by inflation. As each group of workers which considered itself to be earning below-average wages attempted to reach the average, it automatically raised the average itself – which then brought another set of piece rates into the 'below-average' category. Thus wage spiralling, given the system as it was then, was almost inevitable.

Price lists were not the only manipulable element in pieceworkers' wages. Indeed, from 1951, price list revisions had been disallowed, except for a substantial change in working conditions, although the definition of 'a substantial change' was vague. However, when an improvement in the price list could not be negotiated, workers pressed for an upward revision of allowance payments. This largely explains the very high percentage of allowance payments in pieceworkers' wage packets in Yorkshire, to which we have already referred. An earnings norm of 70/- was generally recognised in Yorkshire around 1961. For this money the worker was expected to handfill 14 cwts of coal. But because pit price lists had not been revised since 1951, the filling price of 3/4d per cwt made it impossible for a man to obtain the norm from the price list alone. For this reason it was quite usual for management to pay a consideration payment or 'make-up' to the accepted norm. The consequence of this practice was, as already noted, to considerably weaken the dependence of earnings on output. When tasks increased, the make-up would be reduced, and vice versa. From an examination of tasks and earnings in Yorkshire in 1961, the writer estimated that the average incentive in the price list had been reduced to approximately 1/- per ton. In many collieries an informal minimum operated irrespective of output. In such instances the earnings would only vary when output was above a fairly well-defined level. Thus in most pits there was an accepted minimum of 65/- per shift.

But, while the use of earnings norms may have helped to reduce the disparity in pieceworkers' earnings, disparities were still large, if only because the norms differed widely across the industry, with physical working conditions and variations in local bargaining strength. And such underlying tendencies to earnings equality themselves produced inconsistency in the wage structure as they ignored real differences in performance. Thus men were often paid the same rate for widely different physical inputs and the same physical inputs were rewarded with different rates of pay.

3.3. Job ranking

So far we have concentrated, by way of illustration, on the wages of one occupation. At this point, though, we shall examine the consistency of labour evaluations between different occupations.

All the coalface operations listed in Table 3.1 were placed in the highest underground grade of the national wage structure. One may infer, therefore, that the NCB and NUM thought those jobs were equally skilled and/or arduous. To what extent did actual wage levels reflect that opinion? Table 3.6 gives the rank order of the average EMS paid for each of the five most densely occupied coalface operations (A, B, C, D, E) in the industry as a whole, and each NCB division separately. It also gives the range of EMS between the highest- and lowest-paid of these occupations in each division, and the range of EMS between the highest- and lowest-paid men in the *same* occupational category in the various NCB divisions.

Although for the industry as a whole only 3/10d per shift separated the wages of the highest- and lowest-paid workers, the difference actually varied from 14/5d per shift (in Kent) to 3/4d a shift (in Scotland). And, if we examine the range of earnings for the same job between coalfields (column 11), we find an even larger

Table 3.6. *Mean shift rate (EMS) rankings* of face pieceworkers in 1956[a] NCB Divisions*

Task		(1) G.B.	(2) Scot.	(3) Northl'd Cumbl'd	(4) Durham	(5) Yorks.	(6) Lancs.	(7) East Mids.	(8) West Mids.	(9) South West	(10) South East	(11) Range of E.M.S. s.d.
Conveyor movers	A	1	5	1	1	2	4	3	2	2	1	29 9
Rippers	B	2	3	3	4	1	1	1	1	1	3	26 5
Coal cuttermen	C	3	2	2	3	3	3	2	3	5	4	24 6
Packers	D	4	4	3	2	4	5	4	4	4	2	12 0
Colliers (hand-filling)	E	5	1	5	5	5	2	5	5	3	5	28 9
Range of E.M.S. s.d.	F	3.10	3.4	10.10	8.2	6.4	5.2	4.8	4.4	6.10	14.5	

[a]When two occupations have the same EMS they get the same rank.

Source: Based on information from NCB, Working Papers.

disparity of earnings (from 29/9d for conveyor movers to 12/- for packers).

Large as some of these wage differences were, one might not, for the reasons already discussed, be alarmed if they nevertheless reflected consistent occupational rankings. But no such consistency is apparent in the table. True, rippers, whose job is still thought to be the hardest in the pit, and conveyor movers and coal cuttermen, who managed to command high pay rates with the change to mechanised mining, were generally paid more than colliers and packers. But the ranking of the average wages paid for the occupations listed does not show much more consistency than that. Thus, if one compares the ranking of all five jobs in Scotland and Durham – areas, incidentally, where, on average, seam conditions are equally bad – we find that, whereas in Durham conveyor movers were the highest paid, and colliers the lowest, in Scotland the colliers were the highest paid and the conveyor movers the lowest. Thus, the two districts' ideas on ranking were diametrically opposed.

The overall inconsistency in the pieceworkers' wage structure is reflected in Table 3.7 which gives the rank correlation coefficients of job rankings between the industry as a whole and the various NCB divisions, and between the divisions themselves. Of the 45 rank correlation coefficients shown in Table 3.7, 38 are statistically insignificant at the 5% level, only five are positively significant at 5%, and two are actually negatively significant at the 5% level. It is interesting to note, however, that there is a general tendency for the correlation coefficient to be highest between adjacent areas. Thus, there is a 0.9 correlation between the ranking of face operations in the East and West Midland coalfields, one of 1.0 and 0.9 between Yorkshire and the West Midlands and East Midlands respectively, and one of 0.9 between the rankings of jobs in the Northumberland and Durham coalfields. At the opposite extreme there is a negative correlation of -1.0 between the ranking of jobs in the South East and Scottish coalfields, although one should also note an exception to the general case – such as a positive correlation between the ranking of jobs in the coalfields of Durham and the South East. These figures might suggest some interaction between geographical proximity and subjective job evaluation, while the widespread dispersion of the coalfields was a factor operating against general consistency in job rankings. We shall say much more about regional wage levels, however, in a later chapter.

While the picture traced so far is rather chaotic, it became even more so when a similar exercise was conducted for collieries within particular coalfields. Thus an examination of the distribution of

Table 3.7. *Rank correlation coefficients of job ranking in NCB Divisions in 1956*

	GB	Scot.	N. & C.	Durham	Yorks.	Lancs.	EMs	WMs	SW	SE
Great Britain	1	−0.3	0.8	0.4	0.6	0.1	0.5	0.7	0.2	0.3
Scotland		1	0.75	−0.9	−0.5	0.6	−0.2	−0.5	−0.3	−1.0
Northumberland and Cumberland			1	0.9	0.8	−0.4	0.6	0.7	0.1	0.8
Durham				1	0.3	−0.8	0.1	0.3	−0.1	0.9
Yorkshire					1	0.3	0.9	1.0	0.6	0.5
Lancashire						1	0.4	0.3	0.5	−0.6
E. Midlands							1	0.9	0.3	0.2
W. Midlands								1	0.6	0.5
South West									1	0.3
South East										1

Source: Table 3.6.

individual earnings in 1956 showed that a range of 10/- was needed in the case of each of the jobs listed in Table 3.1 to accommodate the middle 50% of workers and, taking extreme earning levels into account, the scatter of personal earnings in each job was as much as 60/- a shift. When a survey of 338 collieries was conducted, the ranking of jobs for the industry as a whole was repeated in only 2% of collieries.

It would seem, then, that local evaluations of piecework tasks varied enormously in both absolute and relative terms and that these differences were probably as much due to the enormous variations in piecework contracts as they were to differing effort inputs of the workers.

3.4 Reaction to the working party's report

The NCB working party on the pieceworkers' wages system finished its work by the end of 1957. It did not make recommendations as such: rather it presented an analysis of the existing piecerate system and indicated the adjustments necessary if a national piecerate system was to be implemented. Each piecework task was analysed in turn, national task standards were derived from the data so obtained, and computed with actual task performance. Earnings levels for each task were then analysed and set against the ratio of actual performance to the nationally derived task norms, to establish the degree of inconsistency in the wage structure and the extent of the adjustment in piece rates required for consistency. This latter adjustment, however, would ultimately depend on the level at which the national authorities decided to pitch the average level of EMS for the various piecework occupations. Thus, the committee presented, at the cost of much time and money,[7] an analysis upon which a nationally based piecerate system could be derived, which would equate earnings with 'real' tasks performed in any given sets of conditions.

For reasons which no one has been able to explain adequately, the results of the two years work were never presented to the NUM. Of the twenty-one reports submitted to the NCB, only one was actually discussed by an NCB working party established for that purpose. Nevertheless, after 'due consideration' the working party reported to the NUM that the investigation had produced no satisfactory results. Subsequently, the task schedules were circulated to the Divisional Coal Boards, but only to help them in price list negotiations.

One can only guess why the NCB decided not to act upon the reports of its survey team. One factor was probably the feasibility of implementing a revised structure at a reasonable financial cost.

Piece rates would have had to be adjusted, either because, where the existing task norms were thought to be correct, earnings were too high or too low or because, when given a national norm of earnings, task norms were too high or too low. In selecting a national norm of money earnings, the working party tended to favour the average level of the existing wage distribution, on the grounds that this would minimise piecerate adjustments. In the case of colliers, this would have been about £2.55 per shift. But the working party realised that, to be acceptable, a revised wage structure should give as many people wage increases as it did decreases. When this criterion was applied and the proposed task schedules taken into account, a money norm of £2.70 for colliers would have appeared appropriate.[8]

But the implications of setting standard money norms for the industry as a whole would have been very different for different coalfields. Turning, by way of example, to colliers once again, a money norm of £2.70 a shift would have meant, given the existing relationship between tasks and earnings, a fairly general upward revision of piece rates in Scotland and the East Midlands, but relatively few upward revisions in the South East, Northern (N. & C.) and South Wales coalfields. In the South Eastern coalfield, for example, existing piece rates were so out of line that even an earnings norm of £3.0 per shift would have given only 30% of colliers a wage increase, compared with 68% of colliers taking Great Britain as a whole. Thus, once money norms had been set, some areas would have had to increase their task performance considerably if they were not to suffer very sharp wage reductions from the revision of piece rates. In South Wales, for example, an area which already thought itself underpaid by national wage comparisons (although it was not by task standards), it was estimated that an *increase* in task performance of some 30% on average would have been required from the colliers who were not then reaching the national task norm performance, before they could be paid even the earnings norm of £2.70.

The cost of implementing a national piecerate structure would clearly have depended on whether piece rates were to be adjusted immediately to national standards and the level of performance at those revised rates. The writer has no figure for the estimated overall cost of schemes based on the various assumptions that might have been made. In the case of colliers, however, the cost ranged from £0.87 million per annum, assuming an earnings norm of £2.70 per shift and that piece rates were rigorously revised to national standards with no change in the current task performance, to, more realistically, £3.22 million in the first year and £1.66 million in the second year, if above-standard piece rates were allowed to remain in existence for the life of existing coalfaces. But there was no guarantee that the total cost of a piece rate revision (which would have been far greater than the larger of these two estimates for colliers) could have been recouped from increased productivity. There was also the possibility of widespread industrial discontent among pieceworkers, who generally stood to gain much less from the envisaged revision than did daywagemen in 1955, and with whom the NUM officials (many being former pieceworkers) may have had considerable sympathy. Thus the Board's reluctance to act on the working party's reports is understandable.

The Board's rejection of the proposed structure may also have been influenced by the superficial evidence that, in aggregate, pieceworkers' wages were already responding to central direction. The Board had, by this time, begun to make a concerted attempt to coordinate local wages policy and between 1955 and 1957 the rate of growth of pieceworkers' wages was substantially less than it had previously been.

The Board also felt there was little point in revising a payment system which was already being outdated by the spread of power loading. Local wage agreements for power loading already existed in most areas, and it probably seemed more sensible to concentrate on coordinating these agreements than revising those operating on conventional coalfaces. If it could establish general principles to govern the wages of power-loading teams, the conventional pieceworkers' wages structure might be controlled indirectly and gradually disappear as power loading replaced conventional mining. Certainly the Board, at the time, was thinking increasingly of method study as a means of assessing performance.[9] While local custom and practice would almost certainly make this unacceptable on machine-cut faces, the Board thought there was a fair chance of it being accepted as a new system of mining was introduced. Thus, towards the end of 1957, the Board suggested that negotiations with the NUM should start as soon as possible on a national power loading agreement, as the next logical step towards revising the pieceworkers' wage structure.

3.5 Summary and some conclusions

The piecerate system which the nationalised coalmining industry inherited was extremely complex and riddled with anomalies. Overall, there was, for particular jobs, a broad relationship between personal task performance and wage rates. More detailed examination, however, revealed considerable inconsistency, in that widely differing tasks were rewarded at similar wage rates and widely differing wage rates were often paid for similar tasks. While many of these anomalies were probably the product of such historic circumstances as the differing abilities of employers to pay, they were also aggravated by the post-war adoption of comparative earnings norms in local wage bargaining and, associated with this, the manipulation of locally determined allowance payments, which also served to reduce the wage incentive element in the system. Wage differences between jobs were similarly inconsistent, and occupational rankings themselves displayed differing appreciations of a particular job's relative worth, which increased with the distance between coalfields. Thus both economic and institutional pressures were at work to make the wage structure deviate from a common standard.

Cost restrictions and, because of them, fear of industrial unrest, led the NCB to ignore the working party's proposals for piecerate reform presented in 1957, as did the knowledge that the wage system would become increasingly obsolete with the advent of new technologies at the coalface. Subsequently, then, the NCB, which was, in any case, more specifically concerned about the general pace of wage advance, directed its energies to coordinating the district wage agreements which governed power-loading operators, hoping that by so doing they would provide a basis for a national power-loading agreement as the next step in revising the industry's piecerate system.

4. THE REVISION OF THE PIECEWORK WAGE STRUCTURE

In the previous chapter we outlined various features of the industry's piecerate system and suggested reasons why the NCB discarded such ideas as it had for revising it in favour of opening discussions on a National Power Loading Agreement (NPLA) as the next step in revising the piecerate structure. Even so it was nine years before the NCB and NUM could negotiate a suitable NPLA, and another five years before they completed the task of revising the industry's piece-rate structure.

This chapter is divided into three parts. First we examine some of the early difficulties in obtaining a national wage agreement on power loading. Second, we outline the principal features of the NPLA which eventually abolished piece rates at the coalface and set the precedent for the revision of the elsewhere underground pieceworkers' wage structure in 1971, the details of which we outline in the third section.

4.1 Towards a national power loading agreement

4.1.1 *District power loading agreements*

The NUM had asked for a national wage agreement for power-loading operations as early as 1946. On that occasion, though, its demand was turned down by the industry's National Reference Tribunal.[1] Thus, by the time the NCB came around to suggesting a National Power Loading Agreement in 1957 it again had to contend with all the local customs and practices which had evolved in the meantime.

Wage agreements for the earliest machines were usually negotiated at pit level, but by the mid 1950s these had been replaced by district agreements in every coalfield except Lancashire and Kent. There were some important differences between these agreements. The coalfields of Scotland and Durham negotiated daywage agreements which provided for a fixed bonus upon completion of a specific task. All the other coalfields opted for contractual piecerate agreements. In Yorkshire, power-loading teams could choose between a contractual and a daywage agreement.

While these differences in wage practice were an obvious obstacle to getting a standardised NPLA, it was really the NCB's insistence that method study should be uniformly applied to power-loading operations which subsequently proved to be the stumbling block to a national wage agreement. In the absence of a national directive, district and local practice again differed on this all-important point, and it was rather optimistic to expect that it would be universally accepted.

In a few areas – Leicestershire, Nottinghamshire and South Derby – method study of both machines and men had been accepted, if reluctantly, and in North Derbyshire it was being applied unofficially. But in all other areas there was considerable hostility towards method study of either machines or men. Thus feelings at the Union's Annual Conference of 1958[2] ran very high when the NCB's suggestion that tasks on power-loading operations might be assessed by method study was discussed. A Scottish delegate, J. McLean,[3] for example, warned Conference of a national agreement which was designed, in his opinion, to speed up work, make 'robots' of the men, and increase the hazards at the coalface. A Nottinghamshire delegate, J. Tighe,[4] reporting on his area's experience of method study, described it as a 'dismal failure' which 'had created more problems in Nottinghamshire, and more discontent, and come nearer to stopping that great coalfield than even the National Wages Structure'. Another delegate feared that work study would be used by unscrupulous management to depress wage standards and cut costs at times of coal surpluses, while another suggested that, with work study, the machines would set the wage ceilings and these be used to check wages on conventional faces. Not one delegate spoke in favour of the Board's proposal that method study should be applied to both machines and men, and one, S. Unsworth[5] of Lancashire, went so far as to oppose a national wage agreement *per se* as a direct attack on local autonomy. In view of this opposition to the NCB's proposal, the NUM could do little other than inform the Board that tasks should continue to be assessed according to local custom and practice.

There was also strong feeling that any move towards a national agreement on power loading should not be seen as an alternative to revising the existing piecework system and, despite the NCB's opinion to the contrary,[6] the NUM attempted for the next few years to keep the two issues separate.

The deadlock on method study could have been broken in 1959 when, having studied the Board's method study techniques, the NUM executive persuaded its Annual Conference that there was little to object to in method study of the machines so long as it did not apply to the men operating them. However, the Coal Board refused to compromise on this issue and the problem of task assessment prolonged negotiations on an NPLA for a further nine years, during which time the negotiation of such an agreement and the revision of the piecework system became inextricably linked.

Although the NCB and NUM did not agree on an NPLA until 1966, they increasingly used their powers under a national agreement of 1953, which specified that district officials should submit their proposals for revised district wage agreements to the national officials for the latters' approval. Although in most instances the national officials merely rubber stamped district proposals, the NCB, in particular, did take exception to some applications for revision, and was able to introduce greater uniformity into the agreements and their influence on wage trends. We shall have much more to say on this in Chapter 7.

What then were the factors which brought about a reopening of discussions on an NPLA in 1965? During the late 1950s and early 1960s, coalface technology advanced in leaps and bounds. In the early 1950s power loading was held in check because developments in getting and loading had not been matched by similar advances in subsidiary operations, such as roof control, ripping and stable holing (described in Chapter 2), necessary to keep the coalface open. By the 1960s, however, powered supports were overcoming the problem of roof control and hand-got methods of ripping and stable holing were gradually being superseded by machines which made it possible to cut coal on two or even three shifts. The most dramatic advance in face technology, however, was still to come.

Although the techniques described above greatly increased coalface productivity, face technology was still very demanding in manpower. As early as 1960, the NCB had been experimenting with remotely controlled coal-cutting machinery and in 1963 the first Remotely Operated Longwall Coalface (ROLF) cutter started operation in the East Midlands.[7]

Although there were only two remotely controlled installations operating in 1965, the NCB was again forced to rethink its wages policy. A local agreement had already been concluded for the first ROLF installation at Newstead, and there was a danger that custom and practice would again preclude the imposition of a planned wage structure. Once again a new technology was the immediate stimulus to yet another attempt to rationalise the coalface piecework structure.

In November 1965 – three years after the first ROLF installation came into operation – the NCB and NUM concluded a national wage agreement for men operating ROLF installations.[8] Its provisions for men on ROLF faces set an important precedent when, a year later, the NCB and NUM reopened negotiations on a national agreement for power-loaded faces, so it will be as well to outline the main features of the ROLF agreement here.

4.1.2 *Wage agreement for Remotely Operated Longwall Faces*

The ROLF equipment was theoretically designed to eliminate the dependence of face cutting speeds on human effort. In view of this, the NCB decided that it would be inappropriate to adopt a piecerate payment system on ROLF installations. Thus, the most important novelty of the agreement was that it fixed day rates for men operating ROLF installations. The NUM had wanted a uniform national wage rate for these installations,[9] but, in view of existing regional wage district differentials, a uniform national shift rate for ROLF operators was not practical at this time. Thus, the wage district rates varied from £4 per shift (in the East Midlands) to £3.62½ per shift (in other districts). It was agreed that uniformity of rates should be achieved within five years.

Two points need to be noted about the agreed shift rates. First, they bore little relationship to the existing shift rates for face operators in the wage districts. Thus, the £4 per shift agreed for the East Midlands was some 35p per shift *less* than that agreed by the pit umpire at Newstead in 1964. And areas with very different earnings levels, such as Warwickshire and South Wales, were brought onto a common shift rate (£3.62½). In both instances the rate agreed for ROLF teams was less than that which could be earned on more standard power-loading installations. This was justified, it was argued, because the work was much lighter.

Other important features of the agreement concerned the workpeople to be included in the agreement. Despite the technical inference of their title, the first ROLF installations were by no means fully automated. Thus the actual running speed of the machine was technically set, but even with machines the jobs of ripping and stableholing were still largely dependent on physical effort. For this reason they had previously been covered by separate contracts. In the ROLF agreement, however, the new wage rates were to apply to all men engaged on face operations. The rates were thus to apply to men performing the jobs of stable holing and face ripping as well as to the men actually operating the face equipment. The NUM had asked that higher rates be agreed for these jobs where they were unmechanised. The NCB rejected this but decided that the agreement should only operate where the ripping lips and stable holes were mechanised. Second, the agreement provided that craftsmen who maintained the face equipment should be included in the face team and be covered by the agreement. This was a major innovation, and a concession to the union. The Board had always opposed the inclusion of craftsmen in power-loading teams and in no wage district was there a precedent for their inclusion.

Despite the specific circumstances for which the ROLF agreement was negotiated, it formed the precedent for a NPLA in the following year, by which time 80% of coal output was obtained from power-loaded faces.

4.2 The National Power Loading Agreement

The National Power Loading Agreement[10] concluded in June 1966 did not so much revise the predominantly piecework system which previously applied to power-loading operations, as abolish it in favour of a daywage system. The NPLA is clearly a landmark in the history of coalmining wage reform and we shall analyse its effects in detail later. We shall restrict ourselves here to the main features of the agreement.

Because of the very considerable wage differences among power-loading teams under district wage agreements at the time of concluding the NPLA, the NCB could not afford to implement uniform shift rates throughout the coalfields in one step. Thus, the wage rates agreed in June 1966 varied from £3.75 per shift in Scotland, Durham, Cumberland, North Wales, South Staffs and Shropshire, and South Wales, to £4.47 in Kent. These rates reflected the average EMS of power-loading teams in the various wage districts during the six months prior to the agreement. It was agreed, however, that complete wage uniformity between the coalfields would be achieved by 31 December 1971. This was to be achieved by awarding differential wage increases as between the wage districts in the interim. It was initially envisaged, for example, that the wages of power-loading men in Kent – the highest-paid area – would be frozen for a five-year period while the wages of power-loading teams in other areas were raised, at different rates of increase, to the Kent rate.

The NPLA was described as a supplemental agreement to the revision of the daywage structure agreement of 1955 and, like that agreement, the negotiated shift rates were all-inclusive, meaning that they could not be supplemented by locally agreed allowance payments. The only wage elements which could be paid in addition to the national power-loading rates were national payments to first aid men, and payments set out in district wage agreements for carrying flame safety lamps, working in wet conditions and carrying powder canisters.[11] It was also agreed that the team captain should be paid an additional 20p a shift.[12] As from June 1966, the five-day-week bonus, payable to power-loading teams, was to be uniformly 20% of the normal weekly shift earnings.[13] In the case of all workers, except craftsmen included in the team, overtime payments were to be based on the shift rates agreed under the national agreement. Craftsmen tend to work more overtime hours than other workers, and this explains why their overtime premium was to be based on the more unfavourable Underground Grade I shift rate (then £2.40). Guaranteed wage payments, however, were to be based, in every case, on the Underground Grade I shift rate.[14]

Although the NPLA was described as a daywage agreement, it did provide that method study should be used to determine maximum machine utilisation and the number of men needed on each installation on each shift.[15] As such the agreement is better described in principle as one of measured daywork. The manning requirements were to be established initially by observing the working of a team during the first four weeks of operating a new face. The manning during the running-in period was to be fixed by reference to similar installations elsewhere and was to be decided by management, although the agreement did stipulate that the union should be consulted on this question. It was further agreed that the union could have an observer present when method study was being conducted, if it so wished. Should any of the team selected originally prove unsatisfactory, management, in consultation with the union, again had the right to take that person out of the team.[16]

The agreement placed considerable emphasis on maximum machine utilisation, *throughout the full time of each shift*. Furthermore, should utilisation of the machine fall below that which was considered to be maximum, the agreement put team members under an obligation to cooperate with management in ascertaining the reasons for the hold-up or delay and suggest, where possible, ways of overcoming the problem.[17] However, the agreement did *not* stipulate any penalties (such as withholding payment) if this cooperation was not forthcoming.

Under the NPLA, a coalface was defined to include 'any advanced heading or ripping worked in conjunction with the coalface'.[18] Thus, the classification of an F.69 Power-Loader Man, to whom the agreement was to apply, included men engaged on making the face gates or roads and those on stable-hole work, as well as the customary face tasks, such as operating the power-loading machine, advancing the face conveyor and supports, or setting and withdrawing face supports, building the face packs, water infusing and holing and blasting (for water infusion where necessary).[19] This was an important innovation on district agreements – the precedent had been set by the ROLF agreement, concluded under rather different circumstances the previous year. Another provision of the agreement, also traceable to the ROLF agreement, was the decision to include at least one craftsman to be solely responsible for the maintenance of the machine in each power-loading team operating under the NPLA.

All the men included in the power-loading team were to receive the same wage rate, although there were still considerable differences in the physical effort required between the various jobs covered by the agreement. To overcome this problem the agreement stipulated that all team members, other than craftsmen, should be interchangeable between the various tasks which a power-loading team was expected to perform.[20]

Before we leave this sketch of the provisions of the NPLA, attention should be drawn to clause 1 of the agreement, which specified that the agreement should apply only to new faces started on and after 6 June. Thus, faces which were already in operation at that time were not to be subject to the conditions of the new agreement. This meant that until all the faces which had been operating on 6 June had finished and the men transferred to new ones, there might be some faces in a pit operating on the district agreement and others operating under the national agreement.

While the decision to negotiate a national agreement for power-loading operators may be seen as a logical development in wages policy, it is not obvious, in view of earlier negotiations on an NPLA, why the NCB and NUM suddenly opted for a daywage agreement. Indeed, there seems to have been an acceptance, until very late in the negotiations, that the final draft would incorporate a wage incentive system.

The idea of substituting day rates for piece rates was not, of course, new to the NUM; it had been a stated objective of union policy since 1926 and one reiterated in the Miners' Charter of 1946. But as Sid Ford, the then NUM President, said in his address to the Scottish craftsmen as late as 1964,[21] 'We have to face the fact that union policy is, to say the least, confused in that subsequent to 1955 there have been a number of decisions calling for the establishment of a national *piecework* structure. Obviously it is not easy to reconcile these latter decisions with the earlier ones which clearly envisaged the complete elimination of piecework. The fact is, however, that the NUM had been so used to hearing NCB arguments in favour of a national piecework structure *which would encompass power loading*,[22] and so afraid of opposition from its own rank and file, should it go for wage uniformity, that it simply gave up asking for *the abolition of piecework* as a general *policy*.'

The NCB used several arguments to justify a daywage agreement.[23] First, it maintained that technology on power-loading faces had reached the point where piece rates were no longer appropriate. Second, it agreed that daywages would facilitate greater control over faceworkers' earnings. Third, daywages would free management from the time-consuming task of settling wage disputes and encourage greater working flexibility. Fourth, they would encourage greater safety in the pits.

Just when, and how, the NCB adopted these opinions, however, is rather unclear. The first hint of the shift is to be found in the NCB's evidence to the Royal Commission of Trade Unions & Employers Associations,[24] when, in reply to the question from Mr Wigham, 'Would you like to get rid of piecework altogether?', Mr W. J. P. Webber, the Board member for industrial relations, replied, 'Yes, I would like to get rid of piecework, and the more we are mechanising, the less is the effort of the individual involved. We will shortly get to a much more highly mechanical system, and it creates disparities as between the people who are fortunate to be based on a piecework system while there are others to whom you cannot apply it.' Whether Webber had always held that view and it reflected NCB opinion at that time, however, is difficult to assess.

One important factor influencing the NCB towards a daywage agreement, though, was the opinion of the Prices & Incomes Board in their report on coal prices[25] in February 1966: 'We consider that a step has been taken in the right direction by the conclusion recently at national level of an agreement providing for day rates for men operating remotely-controlled machines at the face. Mechanisation, however, makes it desirable that this method of payment should increasingly be substituted for a piecework structure. This is a matter for negotiation between the NCB and the NUM and such negotiations are in train to cover work on power-loading machines. We consider it essential that these are brought to a successful conclusion. The perpetuation of a wages structure which is no longer in keeping with technical developments is hastening the decline of the industry and is at odds with the ambition of the mineworkers' union to ensure fair treatment between man and man.'

At this time the NBPI was particularly concerned about wage drift and its association with payment-by-results systems. And there can be little doubt that the NCB had created the impression that the industry's piecework system was out of control by stating[26] that 'the disparity in earnings between pieceworkers and daywagemen has in fact been widening over the last ten years. Not only is this inequitable; but it also makes for inefficiency and may add to the difficulties in the way of maintaining the labour force at the desired level.'

Wage differentials may be measured in several ways and there is nothing officially to confirm whether the argument was based on absolute or percentage wage differentials and whether the reference was to hourly or weekly wage statistics. The writer has been led to believe (in confidence), though, that the NBPI at least was thinking of percentage shift rate differentials. In fact, as we shall show later, the percentage differential between face pieceworkers and surface daywagemen had narrowed after the later 1950s, if by differing amounts in the various coalfields.

The Board and the union seemed more concerned about the absolute share of the wages cake going to pieceworkers. When the NUM asked for an increase in daywage shift rates, for example, they were reminded constantly that pieceworkers had already taken more than their 'fair slice of the wages cake' and that consequently there was little left for the daywagemen. Just what that meant in practical terms, of course, is extremely subjective: the Board's argument does seem to imply that because technological developments had eased face tasks, the absolute wage differential between faceworkers and daywagemen was too large and would remain so as long as piecework was retained.

There can be little doubt that when the NCB eventually suggested a daywage agreement, the NUM's executive was caught unprepared by what appeared a sudden change of mind. Many of the executive probably thought it had little chance of being accepted in the coalfields anyway. It was thus left to Will Paynter to realise a life-long ambition.

Paynter and his supporters (mostly from the lower-paid areas), knew that, so long as the power to settle wages resided with the district unions and their lodge officials, the national union could never really impose its will over its constituent parts. To Paynter, then, piecework was a direct challenge to the authority of the national union and the national wages policy in which he fervently believed.

But, while this might explain why Will Paynter and other members of the NEC supported the agreement, it was insufficient to have sold the agreement in the coalfields. Equity is all very well in theory but, as we have seen, it has been known to fall on stony ground in practice. In a masterly piece of oratory Paynter stressed the advantages of a planned wage structure to a special conference called to discuss the agreement.[27] But he knew that, ultimately, voting on the agreement would be determined primarily by personal considerations in the areas and here the numbers benefiting from the agreement would be all important.

Although it was the intention of the Coal Board, and apparently of the union as well, that the NPLA should serve to compress wage differentials between faceworkers and daywagemen, the implications of this for future wage trends at the coal face were not made clear to union delegates. The support of the low-paid areas in which faceworkers stood to gain substantial pay increases from the agreement over the next five years was assured, although the spread of pieceworkers' earnings was still such as to guarantee some opposition to the agreement even there. Further support could be assured from the craft unions, as the inclusion of craftsmen in the face teams meant a major breakaway from the national craft structure which had long been a source of complaint. At the vote, the agreement was ratified by a narrow majority of 269,000 to 226,000. The voting was very much as one might have expected. Thus, the relatively high-paid areas, such as Lancashire, Leicester, Midlands, Nottinghamshire, Yorkshire, South Derbyshire and Derbyshire, voted against the agreement, while all the other areas, including the craftsmens' groupings, the cokemen and COSA voted in favour.[28]

The NPLA has been the subject of considerable controversy both inside and outside the mining industry. While the NCB[29] argued that it facilitated greater central control of wages than did the wage agreements it was eventually to replace, and that the abolition of piece rates would improve the climate of industrial relations in the industry, others[30] have contended that power-loading operations were by no means at the level of technical development which warranted a changeover to the daywage payment form and that the cost of the agreement, in terms of lost output, has been considerable. We shall examine the substance of these arguments in later chapters.

4.3 The third national daywage structure

Insofar as the NPLA was to apply only to men classed as F1 power-loading operators, it did not represent a wholesale revision of the pieceworkers' wage structure. The wage levels of pieceworkers operating conventional coal faces were affected indirectly by the NPLA, insofar as the NPLA wage rates were to become new minima below which their wages could not fall. But the wage levels of pieceworkers on conventional coal faces (some 25% of all faces in 1966) which were above the appropriate minima were unaffected. Of course, the number of conventional coal faces was rapidly diminishing by the mid

1960s and the men working them being transferred to power-loading faces, so one could regard the wage structure of their operators as self-abolishing. The same was not true of 'elsewhere-underground' operations. At the end of 1966, the pay of some 70,000 miners, the majority working away from the coal face, was still determined by local negotiations in what Will Paynter[31] characterised as the 'no-man's land, in between the 1955 daywage structure, in most cases, and the power-loader man'.

The problems involved in revising the wage structure of the remaining pieceworkers and taskworkers working underground were exactly the same as they had been in 1957, when the first investigation of the pieceworkers' wage structure was made. First, there was the problem of defining work tasks, given that the jobs of many miners consisted of many components. Second, there was the problem of deciding, in any case, which jobs should be included within any revised wage structure. This problem was compounded by the practice of many pit managers of paying enhanced daywage or task rates for work which was really covered by the 1955 black book (daywage structure): a survey[32] of taskworkers conducted in the South Midlands Area in 1967, for example, revealed that some 2,700 miners were being paid either a task rate or an 'enhanced' day rate, i.e., in the latter case, a rate higher than the standard daywage grade rate but which was not an authentic 'personal' rate. This 2,700 could itself be split in the ratio of 11 to 18 between those men being paid a task rate or enhanced rate for doing a *daywage* rated job and those paid a task rate for doing a task rated job. Third, there was again the inevitable problem of cost involved in any wage revision, this being exacerbated by the very large wage dispersion among men doing similar work. This problem is well illustrated by Figure 4.1, which shows the average shift rates, in April 1966, among Grade I rippers, who were numerically the most important occupational group among the remaining underground pieceworkers and taskworkers.

Rippers working in the roadway headings were on a par, in the national grading structure, with the majority of faceworkers and were thus subject to the Grade I (underground) standard grade rate. Quite clearly, however, the rippers' EMS depicted in Figure 4.1 bore little or no relation to the standard national grade rate, nor did it display any consistency between pits. The modal wage rate was between 64/- and 68/- per shift, but some men received as little as the standard grade wage rate and others well over 100/- per shift. The highest EMS paid was 160/- per shift.

The wage distribution of rippers was rather larger than that of many other occupational groups to be included in the revised struc-

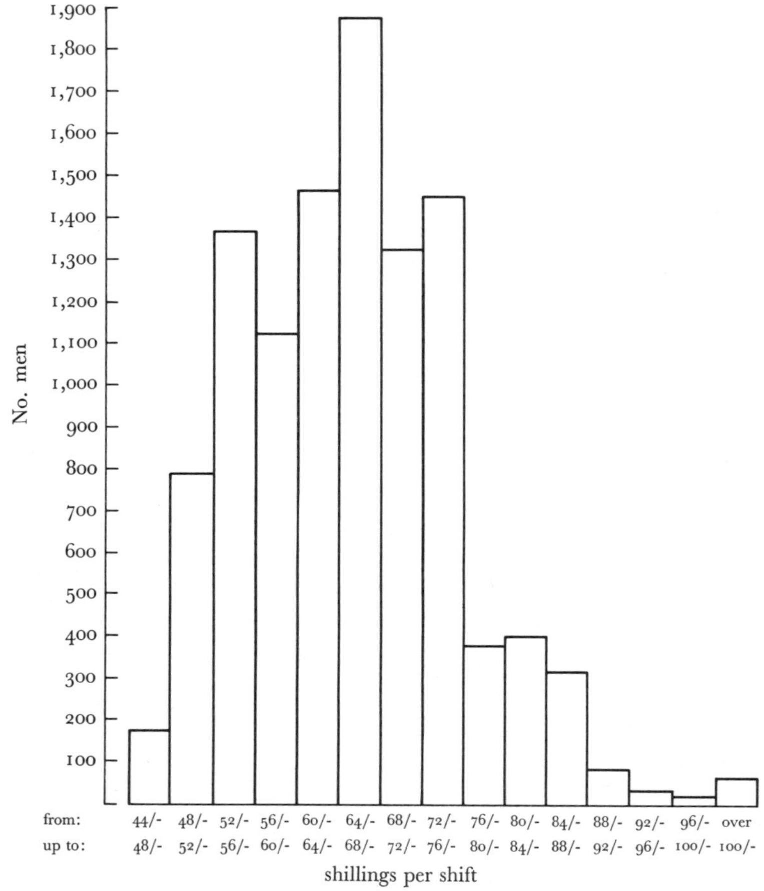

Figure 4.1 Distribution of 'elsewhere underground' rippers, April 1966 (sample size: 10,958)
Source: NCB

ture. Even so, it would have required a considerable sum of money to bring the remaining pieceworkers and taskworkers on to a daywage rate which, in the case of such men, would have had to compare favourably, if not identically, with the district power-loading rates agreed in 1966. This, and the other problems mentioned, all pointed to a fairly lengthy period of investigation, collation and negotiation before agreements would be reached on a wage structure for the remaining pieceworkers and taskworkers. While this was going on, however, steps were taken to accustom the remaining pieceworkers and taskworkers to the idea of national control by restraining their rate of wage advance and reducing the existing wage disparities.

The first step in the long haul towards greater wage uniformity was made in March 1969, when it was agreed that henceforth all adjustments to the wages of the remaining pieceworkers and taskworkers were to be allowable only by national agreement.[33] To this end increases were to be based, in the case of pieceworkers, on the average shift rate calculated over the four weeks as from the pay week beginning 13 October 1969, and, in the case of taskworkers, the task rate applicable on 1 January 1969.[34]

To facilitate the desired narrowing of the wage dispersion, the national wage increase, which was to take the form of flat-rate additions to the consolidated shift rates, was to be scaled according to the level of the prevailing rates in the manner depicted in Table 4.1. Having agreed the addition, however, earnings would still vary in accordance with the existing contract.

It was further stipulated that no increase should be of such an amount as to raise the shift rate of men already earning an average of 90/- or more, and that no increase should be awarded to any pieceworker or taskworker who had benefited from a wage increase during the previous twelve months, or where a shift rate was clearly high in comparison with other rates paid for similar work. This latter clause was very ambiguous and necessarily difficult to implement.

The agreement of March 1969 was followed, the next year, by two similar agreements – one in January and the other in December – while negotiations were proceeding on a revised wage structure. In these the national wage increases were more heavily biased in favour of the relatively lower-paid pieceworkers and taskworkers than had been the case in the 1969 agreement. The agreements also took cognisance of the differences which had developed in the payment of the five-day-week bonus to the remaining pieceworkers and taskworkers.

Since its inception in 1947, the five-day-week bonus was usually paid at the rate of 20% to daywage workers and at 16% to

Table 4.1. *National additions to the consolidated shift rate of elsewhere underground pieceworkers and taskworkers as at March 1969*

Consolidated shift rate	Increase per shift worked
up to 74/-	2/-
over 74/- and under 78/-	1/8
over 78/- and under 82/-	1/4
over 82/- and under 90/-	7d
over 90/-	nil

Source: NCB, *Memoranda of Agreements*, 1969.

pieceworkers. The application of this ruling was fairly easy to interpret except in the case of taskworkers, some of whom were paid what were essentially enhanced day rates and others fixed task rates plus a completion of task bonus. There had never been a ruling as to whether taskworkers would be paid a bonus of only 16% (like pieceworkers) or of 20% (like daywage workers). In the district power-loading agreements of Scotland and Durham, however, where power-loading men were conditioned to task rates plus a completion of task bonus, a five-day-week bonus of 20% had been paid. Although the majority of pieceworkers were found to be in receipt of a five-day-week bonus of 16%, some taskworkers were on a bonus of 16% whereas others received a bonus of 20%. So as to ensure equity of treatment in respect of wage increases, a distinction was made in the schedule of national wage increases in the agreement of 1970 between workers in receipt of 16% and 20% five-day-week bonuses. The rates of wage increases awarded by the agreement of January 1970 are given in Table 4.2.

In addition to stipulating a much more progressive scale of wage increases – from 4/- a shift to zero – compared with the 1969 agreement, the agreement of January 1970 raised the level for which the maximum wage increase was permissible quite considerably, and compressed the range of pay over which some form of increase could be given. It can be presumed, therefore, that this and the subsequent agreement facilitated a fairly substantial contraction of the wage structure of the remaining pieceworkers and taskworkers. In the case of rippers, for example, the quartile deviation of shift rates was some 25% (or 16/-) by the beginning of 1971, compared with 33% in 1966.[35]

Table 4.2. *National additions to the consolidated shift rates of pieceworkers and taskworkers as at January 1970*[36]

Shift rate at Jan. 1970		Shift rate at Jan. 1970	
Rates attracting 20% five-day-week bonus	Increase per shift	Rates attracting 16% five-day-week bonus	Increase per shift
up to 86/9	4/-	up to 89/9	4/2
86/10–88/9	3/4	89/10–91/10	3/5
88/10–90/7	1/6	91/11–93/8	1/7
90/8–93/-	1/-	93/9–96/2	1/-
Over 93/-	nil	Over 96/2	nil

Source: NCB, *Memorandum of Agreements,* 1970.

The last stage in the development of a new daywage structure came in December 1970. This agreement was relatively simple, in that it awarded an increase of 7/11d per shift to workers getting the 20% bonus but less than 100/- per shift basic rates, and 8/2d per shift to workers in receipt of 16% attendance bonus so long as their consolidated shift rate was not in excess of 103/5d. Workers who were already in receipt of shift rates in excess of these amounts received nothing.

While these phasing agreements were successfully contracting the wage structure to a point where the NCB and NUM could sensibly discuss the fixing of national standard wage rates in respect of the jobs which were to be included in the wage structure, a joint working party had been hard at work. Negotiations on a revised structure proceeded rapidly during 1971 and by June the two sides had agreed upon a suitable format. The proposed wage structure was accepted by a delegate conference of the NUM by a vote of 67 to 22 (on the block vote system) and subsequently by the miners (at a pit-head ballot) by 240,000 to 73,000.

The new daywage structure for hitherto pieceworkers and taskworkers was to consist of three grades – A, B and C. A detailed picture of the allocation of jobs between these three grades can be obtained from Appendix 4.1. Broadly speaking, however, such jobs as ripping and filling on non power-loading faces, withdrawing equipment on power-loading faces, sinking and widening shafts and roadway ripping were placed in Grade A. Preparatory jobs, such as the cutting of drivages, drifts, headings and coalfaces and repairing work behind the coalface, were placed in Grade B, whereas most of the remaining jobs, such as supplies transporting, track laying and spillage clearance, were placed in Grade C.

It was agreed that the men performing jobs defined in Grade A would be paid the same as the NPLA rate which, at that time, differed between the coalfields but would be uniform by the end of 1971. This gave Grade A men a weekly minimum of £30. Rates for Grades B and C were fixed by taking the average wage paid for all the jobs in each grade throughout the country. Grade B men were to get £4.1452 a shift, or £24.925 a week and Grade C men, £3.8667 a shift, or £23.20 a week.

Two major problems had to be solved in implementing the agreement. The first concerned the absorption of genuine pieceworkers and taskworkers into the new structure, given their existing diversity of wage rates. The second concerned men who, although classed locally as taskworkers, were really performing a daywage defined job under the 1955 grading structure.

When considering the first problem, it was decided initially to treat pieceworkers and taskworkers differently insofar as phasing was concerned. In the case of taskworkers, it was agreed that they should come within the new agreement immediately. Thus, if a man's rate was lower than the third structure rate for his job, he would be brought up to the new grade rate immediately. Any person getting more than the new grade rate, however, would keep that as a 'personal' rate until he ceased to be employed, or work on the task for which the rate was given, or until such time as his personal rate was overtaken by increases in the national shift rates.

It was decided initially not to bring pieceworkers into the new structure until NPLA parity had been achieved between the coalfields, which would have been in December 1971. It was subsequently appreciated, however, that in the case of pieceworkers receiving *less* than the new daywage rates, this would be anomalous compared with the treatment of taskworkers in the same wage position. Thus, it was agreed to bring such pieceworkers onto the Grade A rate immediately, together with any pieceworkers engaged on contracts which terminated before NPLA parity in December and who on average – over four typical weeks – earned less than the grade rate. After NPLA parity the remaining pieceworkers were to retain their existing contracts until they were transferred to 'new work'. Thereafter, they would be paid the new rate for the job under the agreement. Even so, whereas taskworkers on rates *above* the new schedule kept those rates on a 'personal' basis, pieceworkers in receipt of average shift rates above those in the schedule were brought immediately on to the new rates and thus suffered a wage reduction.

So far as the second problem was concerned, the NUM in Nottinghamshire had already promised militant action if the Board and union agreed nationally to downgrade their 'pseudo' taskworkers to the day rates payable under the 1955 agreement. To get round this problem it was agreed that, although 'pseudo' taskworkers *would* subsequently be covered by the 1955 agreement, rather than by the new one, their rates, if higher than those specified in the 1955 agreement, would continue to be paid on a personal basis. This treatment of taskworkers and enhanced daywagemen meant that, like the phasing of daywagemen into their agreement, personal rates would continue to be paid for some time into the distant future. It was agreed, however, that the whole question of piecework contracts still in existence, and taskworkers still on personal rates, should come up for review in April 1974.

The conclusion of the 1971 Agreement on the underground pieceworkers' structure marked the last stage in the process of transferring the coalmining industry's wage structure to a national basis and establishing uniform standard rates of pay in the industry. As we shall see in the next chapter, however, it did not mark the end of wage reform. No one would expect a national wage structure to go unrevised. Indeed, one test of the efficiency of a national wages policy must be its adaptability to changing circumstances. Both the 1966 and 1971 agreements necessitated the national authorities changing the wage structure of other groups, as did changes in technology in the interim.

4.4 Summary

Although the NCB hoped to conclude an NPLA in the late 1950s as the next step in revising the pieceworkers' wage structure, opposition to method study within the NUM proved a major stumbling block, as did the reluctance of area unions to surrender local autonomy in wage matters. Technological developments provided an opportunity for a major breakthrough in 1965 and, under pressure from the NBPI, the NCB seems to have been persuaded to abolish rather than revise the piecework system in 1966. The ROLF and NPLA agreements thus set a precedent for the extension of uniform day wages to 'elsewhere underground' piecework tasks in 1971. Support for uniform daywage rates in the case of power-loading operators was by no means unanimous however: without the element of surprise, the political and oratorical skills of Will Paynter, and the support of minority groups in the NUM, it is doubtful if they would have been accepted at that time.

APPENDIX 1 THE ALLOCATION OF JOBS IN THE THIRD DAYWAGE STRUCTURE

Jobs in the third daywage structure are divided into three grades as follows:

GRADE A

On drivages, drifts, headings or coalfaces (other than as a power-loading team member): cutting, getting and filling, setting and withdrawing supports, building packs, grading and supporting faults, advancing face conveyors, making gates or roads, and drilling for blasting, water infusion or strata control.

On power-loaded faces: grading and supporting faults, and installing or withdrawing equipment.

In shafts: sinking, widening, and major repairs to shaft linings.

Elsewhere: systematically enlarging a roadway by ripping, dinting, removing or replacing supports, and making large excavations for roadway junctions, engine houses, air crossings, overcasts and bunkers.

GRADE B

On drivages, drifts, headings or coalfaces: preparing for cutting, where not part of cutting-team work; operating water infusion equipment; and face repairs including dinting.

Elsewhere: operating ROLF face consoles, installing or withdrawing equipment on non-power-loaded faces, withdrawing supports and equipment from roadways, roadway repairs, boring for water, gas, drainage or exploration, conveyor or stage loader extension or dismantling, shuttle car driving for mineral clearance, and sealing against heating.

GRADE C

On drivages, drifts, headings or coalfaces: cleaning up spillage, checking supports and recovering buried supports, and moving up electrical switchgear and power packs.

Elsewhere: laying loco rail tracks or track weighing 40 lb a yard or more, installing mono-rail for man-riding, transporting and handling heavy plant, transporting and handling supplies to headings or faces.

NCB, *Coal News*, July 1971

5. THE EVOLUTION OF THE DAYWAGE STRUCTURE AFTER 1955

So far we have traced the various stages whereby the NCB and NUM evolved a national wage structure incorporating the principle that men performing similar types of work should be paid the same rate. In the course of the preceding chapters we have illustrated many of the difficulties inherent in wage structure reform, which favoured a gradual and rather pragmatic approach to the problem. We showed, for example, that, whereas the NUM was officially committed to wage uniformity, its approach had nevertheless to recognise local loyalties and jealously preserved autonomy. Nor could the NCB ignore the threat which too rapid an alteration of the wage structure might pose to coal production, or the very real financial cost which a reform of the structure would entail.

But while many factors favoured gradualism, would not the very same difficulties also make for problems in regulating the revised structures? Would not the national negotiations have to take cognisance of the obvious dissatisfaction displayed by important sections of the mining labour force who were expected to bear the cost of the reforms, or run the risk of losing control of the wage structure? And would they not have to pay lip service to the changing importance of occupational groups within the wage structure or run the same risk? The wage reforms had themselves created anomalies between one work group and another which, left uncorrected, could have posed a serious threat to the efficient working of the industry. Before going on to examine the extent to which wage trends and movements actually reflected national wage decisions, therefore, we shall focus in this chapter on developments in the daywage structure after 1955, to see how the NCB and NUM attempted to meet some of the problems outlined above at a national level.

5.1 The 'rate for the job'

The underlying principle of the 1955 agreement was to pay the 'rate for the job' irrespective of the part of the country in which that job was performed. The dispersion of wage rates among daywagemen at that time, however, was such that it was impossible to fix standard rates at a level which would give complete wage uniformity without some men having their rate reduced. This problem was avoided by allowing men on wage rates above the agreed standards to retain them on a personal basis.

We noted in Chapter 2 that, if the principles of the daywage agreement had been adhered to, it should have been possible, on realistic assumptions, to achieve wage uniformity in three annual wage awards. The ability of the Board to pursue this policy was soon tested.

In the first round of wage negotiations following the introduction of the new structure, the NUM, under considerable pressure from the Midlands Areas where some 70% of 'mark-time' men were concentrated, asked that the ceilings be revised so as to give a larger proportion of men a wage increase than might otherwise have been the case. 'We recognised' said the Executive Committee in its Annual Report to Conference in 1956[1] 'when the wages structure agreement was introduced, that a wage structure in the sense in which we understood it would not be achieved in one step. We were also forced to the conclusion that even with the present increase in wages, we had not yet fully realised the aim of equal wages for equal work for daywagemen. There were still in all coalfields a number of men in receipt of personal rates who would not receive the whole of the increase of 12p or 14p per shift as the case may be. We decided to increase the limits for certain grades where this problem appeared to be most acute.' This quote is not taken out of context! Rather it illustrates the equivocal stand which the NUM was taking on equal pay.

In 1957 the union went even further and submitted a claim for a *general* wage increase *completely free of wage ceiling constraint.* This was objected to by the Board as 'a fundamental attack on the wage structure'.[2] But, while the union's request was turned down, the Board conceded that Grade I underground workers on personal rates above the ceiling of £2.00 should receive some increase so long as it did not take their rates over £2.12½, and that *all* other workers should receive an increase so long as it did not take their rates above £2.02½, an amount which for most grades was considerably above the old ceilings. Some effects of this wage agreement are calculated in Table 5.1.[3] Thus, column I of the table shows the percentage of men on 1956 standard rates and therefore entitled to a full wage increase in 1957. Column II shows the percentage of personal rate men who would have received only a part of the 1957 award had the 1956 ceilings been held. Column III shows the percentage of 'mark-time' men who would not have benefited from an award in 1957 if the ceilings had been held. In column IV we have the percentage of 1956 'mark-time' men at each place of work who actually benefited from the 1957 award and column V shows the number of 'mark-time' men who remained 'marking-time' after the 1957 award as a percentage of all daywagemen.

Although the Board resisted a general wage increase, the union did, in fact, obtain a very good second best in that some 85% (12,500) of the industry's 'mark-time' men received some increase. The award did have the effect of bringing all but 1.0% of the industry's daywagemen within the agreed wage ceilings. But, by actually raising the ceilings, the 1957 award also had the effect of giving a large proportion of personal rate men a full wage increase in 1957, and made a greater proportion of personal rate men eligible for increases in the future. The agreement thus considerably postponed the time at which 'personal rate' men would be put onto standard wage rates. The 1957 award, then, may be considered a partial breakdown in structural policy with respect to 'personal rates' under pressure from the 'higher-paid' areas.

A comparison of Column I of Table 5.1 with data on the spread of day rates in 1955 (Fig 2.2) shows that the percentage of non-craftsmen on standard wage rates increased from 76% to 82% between 1955 and 1956. This process was repeated at each place of work and particularly at the coalface. The writer has estimated[4] that

Table 5.1. *'Mark-time' men and the 1957 pay award*

	Percentage of men				
	I	II	III	IV	V
Place of work	On standard rates 1956	Above standard but below 1956 ceiling rates in 1956	'Marking-time' on and over 1956 ceiling rates in 1956	1956 'Mark-time' men getting increases 1957	'Mark-time' men after 1957 award
Coalface	72	12	16	63	6.0
Elsewhere underground	82	12	6	85	0.5
Surface	86	8	6	99	0.1
Total	82	12	6	85	1.0

Source: NCB, *Wage Surveys*, 1956–58

after the 1957 award, some 88% of daywagemen were on the standard grade rates. Table 5.2, which gives the percentage of men on and above the standard rates in 10p wage bands at each place of work, shows that by 1966 some 95% of daywagemen were on standard rates and only 0.4% were 'marking-time'.

Perhaps the most interesting aspect of the 1966 personal rate distribution is that the highest proportion of personal rates was among surface daywagemen, whereas in 1955 and 1957 this group contained the fewest. This is particularly difficult to explain given that, in *1955*, no more than 1% of daywagemen were on rates more than 30p above the standard grade rates and that standard rates had been increased by over 75p since then. Thus, it was just in that area where, given the frequency distribution of personal rates in 1955, it should have been easiest to achieve wage rate uniformity that there was least progress. From this it may be inferred that, after 1957, the diminution of men on personal rates was achieved through men changing their jobs, or moving pits, rather than through the application of the wage ceilings, which were, in any case, raised as a matter of course in successive negotiations between 1957 and 1966. Further, one may conclude that there had been some 'engineering' of personal rates at local level, so that many personal rate men retained their wage advantage and/or additional personal rates had been created. We shall return to this in later chapters.

Table 5.2. *Adult daywagemen paid standard and personal wage rates, April 1966*

	Percentage of daywagemen on:								
		Personal rates (range above standard)							
	Standard Rate	1–9p	10–19p	20–29p	30–39p	40–49p	50p and over	Total	% men on personal rates over the wage ceiling
Face %	94.7	2.0	0.5	2.1	0.6	—	—	5.3	0.6
Elsewhere underground	95.6	3.0	0.8	0.4	0.2	—	—	4.4	0.6
Surface %	94.4	3.8	1.7	—	—	—	—	5.6	0.1
Total %	95.3	3.1	1.2	0.3	0.1	—	—	4.7	0.4

Source: NCB, Survey of Personal Rates, 1965

5.2 Allowances

By consolidating and then abolishing all the various bonus and allowance payments which had previously been an important, but complicating, element in the wage packet of daywagemen, the 1955 daywage agreement greatly simplified the wage system. Thereafter the rates agreed were inclusive of all bonuses, flat rates and allowances – the only exceptions being the attendance bonus, payment for working in wet conditions and rent allowances. This was a drastic departure from previous custom and practice, and pressure soon developed for a restoration of allowance payments to cater for the varied conditions under which pit work was performed. Thus, the NUM's National Executive Committee reported to their Annual Conference in 1956 that 'one of the most difficult problems we have had to face since the acceptance of the wage structure agreement is that of allowances'.[5]

The first attempt by the union to extend the scope of allowance payments was made soon after the 1955 Agreement was introduced, following submission of many instances in which the areas thought extra payments justified. Thus in August 1955[6] the union presented the Board with a claim that additional payments be made in respect of the following: working in dust – underground and on the surface – working in heat, carrying ambulance boxes, having first-aid responsibilities, carrying powder tins, carrying ring-rose gas detectors, working on dirt heaps, working in fumes or at abnormal heights, being called from home during leisure hours (standby payment), emptying chemical buckets, and carrying injured workmen outbye.

The Board pointed out that the 1955 agreement had been a joint endeavour to eliminate the need for a system of allowances which had grown up in the industry. Nevertheless, while it rejected the general claim as 'a case of nibbling at the wage structure',[7] it did agree to look into the case of first aid men and men called from their homes in leisure hours. Here the position remained until 1959, when the National Reference Tribunal was brought into the picture.

In arguing before the Tribunal[8] against the union's claim for allowances, the Board emphasised that all factors affecting jobs were fully taken into account when they were graded, and that there had been no change in circumstances which would justify a departure from the basic principle of an all-inclusive rate for the job. The Board stressed also that allowances were extremely difficult to distribute fairly and control, and that the granting of some allowances would encourage demands for more. In the Board's opinion, a reintroduction of allowances would be a retrograde step tending towards a reintroduction of the inconsistencies that previously existed, and would involve a negation of the fundamental principle of the daywage structure under which an all-inclusive rate was negotiated.

While the union said it was against returning to the previously irrational wages structure, and *was opposed in principle to have wage determinations at a pit, area or division level*, it claimed that allowances would pose no threat so long as they were agreed nationally. An important part of the NUM's case rested on comparability, first with pieceworkers, for whom in their unrevised structures there was ample provision for allowances, and second with other industries, where provision was often made for allowances in the case of men working in abnormal conditions.

These arguments are extremely interesting. They show that while the union was now committed in principle to a nationally determined wage structure, it could not, so long as some workers had their wages regulated locally, ignore the pressures which these local assessments created. It also illustrates the problems that confront an industry should it attempt, in isolation, to rationalise its wage structure.

The Tribunal found that the payment of allowances *was* consistent with the daywage agreement if they were for exceptional circumstances and only temporary. Furthermore, the Tribunal thought that the Board had set its own precedent in this matter by having already negotiated extra payments (such as for working at heights) in March 1959.[9] The Tribunal recognised, however, that in some instances there would be difficulties of demarcation between normal and exceptional conditions, but thought that 'given goodwill on both sides these difficulties would not seem to be insuperable'.[10]

The decision of the National Reference Tribunal was followed in 1961 by an agreement specifying allowance payments for working at height, or in dust, heat, noxious fumes and water. These allowances payments have since been increased at frequent intervals, and others have since been added to the list.

All these *allowances* are standard, although the *circumstances* which justify them can only be determined by local haggling. But, while the National Reference Tribunal may have held a surprisingly unrealistic view of the day-to-day wranglings which inevitably go on in a pit, it is difficult to see how all the variations in conditions for which they were meant to cater could have been adequately allowed for in a standard wage-rate structure.

5.3 THE WAGE RATE AND GRADING STRUCTURE 1955–74

Once the daywage structure had been implemented, the NUM could seek to increase the wages of daywagemen in either of two ways (or both simultaneously). First, it could negotiate a revision of standard rates and wage ceilings (and also allowances). Second, it could negotiate on the grading of occupations.

We have already examined the main features of wages policy in relation to wage ceilings and touched on the problem of allowances. We shall return to the latter topic when we examine the craftsmen's structure. For the present, let us examine policy in relation to standard grade rates over the period 1955–74.

Table 5.3. *Standard shift rate differentials under 1955 daywage agreement 1955–74*

	April 1955	June 1966	Nov. 1971	March 1974
Grade	Percentage differentials in standard wage rate			
Non-craftsmen				
Surface H/L	112.2 (127.4)+	108.3 (126.8)+	103.5 (119.3)+	102.5 (113.8)+
Underground H/L	114.4	108.5	103.5	102.4
Overall HU/LS	130.2	117.4	112.4	115.2
Craftsmen				
Surface H/L	118.4	125.4	117.5	112.5
Underground H/L	116.4	123.3	116.1 (135.8)*	118.1 (123.6)*
Overall H/L	130.9	133.4	126.3 (147.8)*	132.9 (139.0)*
Craft/non-craftsmen				
Surface HC/LNC	131.0	131.3	115.2	113.9
Underground HC/LNC	127.2	129.1	118.0 (138.0)*	119.4 (122.0)*
Overall HC/LNC	144.9	140.0	128.0 (150.0)*	134.3 (140.6)*
Surface HC/HNC	116.7	121.2	115.2	111.0
Underground HC/HNC	111.8	118.9	114.0 (133.0)*	116.6 (122.0)*

H = Highest grade
L = Lowest grade
NC = Non-craft U = Underground
C = Craft S = Surface
+ = including Grade 1A winding enginemen
* = including Grade 1A craftsmen in face teams
Source: NCB, *Memorandum of Agreements*, 1955–74

Table 5.3 shows the differential between the standard shift rates of various grades in the daywage structure between 1955 and 1974. In each case the grades selected are those at the top and bottom of the grading structure for each place of work. The differential expresses the higher of the two grade rates as a percentage of the lowest.

When the daywage structure was negotiated in 1955, the Board and union agreed upon very small monetary differentials between the various grades. Thus, in an article written shortly after the structure was introduced, two of the Coal Board's 'revisionists'[11] declared: 'The model introduced in April, 1955, may not be ideal, but the structure is adaptable and will no doubt be improved with time. *Consideration of cost*[12] prohibited the immediate introduction of more substantial differentials between standard rates. Even slight changes in differentials involved heavy cost because of the concentration of men at the lower end of the wage spectrum of each grade. The immediate introduction of more substantial differentials would also have meant greater disturbance of wage relativity, possibly to the point of explosion.'

But if these were the reasons for the small initial differentials, do they explain the observed trends in wage differentials after 1955? Table 5.4 shows that the percentage differential declined quite sub-

Table 5.4. *Allocation of non-craft daywagemen in the national grading structure 1955–74*

	1955 %		1966 %		1974 %
Underground					
Grade I	13.6		10.5		8.7
II	8.6		19.2		16.3
III	5.3		7.5	70.2	75.0
IV	51.8	72.5	62.7		
V	20.7				
	100		100		100
Surface					
Grade IA	4.0		4.5		4.8
IB			2.3		3.9
I	5.4		7.3		9.7
II	20.2		27.6	85.9	81.8
III	27.6	70.4	58.3		
IV	42.8				
	100		100		100

Source: National Coal Board.

stantially between grades of non-craft daywagemen. Only between craftsmen and non-craftsmen before 1966, and between the various grades of craftsmen at each place of work over the same period, did the wage differentials increase in percentage terms. Let us consider these various aspects of national wages policy in turn, concentrating first on the non-craft daywagemen's structure.

If finance had been available, the NCB planners would have liked to increase the percentage differentials between the various grades in the 1955 daywage structure so as to give them more meaning – there being only 20p between the highest and lowest non-craft underground grade and only 15p between those grades at the surface. The Board did try to negotiate a percentage wage increase in 1957.[13] But, apart from agreeing that the very highest grade of surface and underground worker should get an extra $\frac{1}{2}$p per shift compared with other non-craft daywagemen, the NUM preferred to settle for a uniform flat-rate addition for all other grades. And, on the NUM's initiative, this pattern was repeated in every wage round before 1974.

Several factors seem to have encouraged the union to go for flat-rate increases.[14] First, the industry had had a fairly long history of flat-rate awards before 1955 and the tradition, no doubt, persisted. Second, the 1955 agreement was based on job evaluation (of a sort). Although it implies an odd kind of money illusion, the union thought it would be difficult to tell a man one day that his job was now worth *x* pence less than that of another man's, as implied in a percentage wages policy. Third, the union was, nevertheless, anxious to alter the *gradings* of many of the lowest-paid jobs and it appreciated (as did the Board) that this would be more acceptable the smaller were the differentials between the grades. Fourth, the union was still trying to narrow the wage gap between face pieceworkers and daywagemen. Although face pieceworkers were only subject to the national grade rate in exceptional circumstances, they were, it was reasoned, conscious of the increase in that rate when negotiating new price lists. Thus, given that any alteration of the existing wage differentials could be to the advantage of only the highest grade workers, so small were the differentials, the union was reluctant to ask for percentage increases in the standard shift rates.

Whatever the rationale of these arguments, the wage differentials between non-craft daywagemen became almost completely meaningless, thus removing a considerable amount of incentive from the wages structure. Between 1955 and 1974, for example, the differential between the highest and lowest grade of non-craft surface workers declined from 112.2% (if we exclude winding-enginemen who had a higher grade to themselves) to 102.5%, and the differential between the highest underground non-craft grade and the lowest surface non-craft grade declined from 130% to 115%. In 1974 the first steps were taken to widen the differential between underground and surface workers – more on this after we have considered actual wage movements.

The differential contraction was also helped by the NUM and NCB agreeing to 'roll-up' the grade structure. In 1964, grades V underground and IV at the surface were abolished – the men affected being upgraded to the immediately higher grade, and in 1970 grades IV underground and III at the surface were also 'rolled-up'. While both these changes were made to help the lower paid, their effect was to prevent any real change in job content being allowed for by wage-rate adjustments. This helps to explain why more devious means were found to compensate daywagemen for such changes. Table 5.4, for example, shows that in 1974 approximately 75.0% of daywagemen working elsewhere underground were in grade III, whereas 81.8% of surface daywagemen were in surface grade II. Over the period 1955–74, there has, with the exception of grade I underground, been a proportionate increase in the number of jobs in the highest grades. This is due to the increased importance and more technical nature of some jobs, such as those involved with the haulage system. But it is also the consequence of pressure from the NUM to upgrade jobs which could not be considered on a par with the jobs being upgraded in the 'rolling-up' process. The elimination of the two lowest grades in 1970 caused particular resentment – mainly among haulage and vehicle drivers – with whom the previously lower-graded men were placed on a par. This led the NCB and NUM to negotiate a separate four grade structure for vehicle drivers in August 1972.

5.4 The craftsmen's wage structure

We have seen already that the standard grade-rate differential between craft and non-craft daywagemen moved decidedly in favour of the former between 1955 and 1966. This reflected the industry's increased reliance on craft skills as a result of power loading and technical advances in ancillary services, such as haulage, which accompanied it, at a time when skills were in short supply. While the colliery labour force as a whole declined from 571,000 to 382,500 between 1955 and 1966, the number of craftsmen employed increased from 51,500 to 56,900 and the proportion of craftsmen in the colliery labour force increased from 9% to 15%.[15] These trends strengthened the craftsmen's bargaining position within the NUM, which became more sensitive to the accusation that their interests were being sacrificed for the benefit of face pieceworkers, who operated but could not maintain the face machinery.

We noted in Chapter 3 that craftsmen were dissatisfied with their relative position in the grading and standard wage-rate structure of 1955. By 1959, the NUM was under considerable pressure to look after the interests of this growing minority. Thus, in a resolution proposed by Group No. 2 in 1958,[16] the National Executive Committee was instructed to secure a substantial improvement in the pay position of craftsmen relative to other workers in the industry. At this time, the immediate pressures for a review of the craftsmen's position seem to have arisen from craftsmen leaving the industry, long hours of work and the imminent introduction of the condition that a craftsman would have to obtain a Class I certificate before being in charge as a mechanic or electrician in the pits.[17] Some additional payment for responsibility was given in 1959, but the NUM felt it too small an inducement to ensure a steady flow of young recruits and compensate craftsmen for the additional skills they were expected to acquire.

As these pressures mounted, craftsmen enjoyed a steady improvement in their absolute national wage-rate differential over non-craftsmen. Furthermore, Grade I craftsmen, because of the relative increase in demand for their skills and the need for them to acquire extra ones, successfully improved their percentage wage differential over their Grade II colleagues.

Nor was the relative improvement in the craftsmen's position in the national structure up to the mid 1960s restricted to their getting larger standard wage-rate increases than non-craftsmen. Although they are not accounted for in Table 5.3, craftsmen also benefited quite considerably from extra allowance payments negotiated at national level. We have mentioned already the special allowance given in 1959 for responsibility. This had risen to 25p a shift in 1961, when another 50p a week was also given to mechanics and electricians with special statutory responsibilities. And later in that year, a further $12\frac{1}{2}$p per shift was agreed for craftsmen who spend more than $1\frac{1}{2}$ hours at the coalface. With new regulations governing the flexibility of craftsmen in 1965, however, it was agreed to consolidate the 25p responsibility allowance into the shift rates of colliery fitters and electricians in three stages.

These allowance payments complicated the craftsmen's wage structure and, while the consolidation of 1965 helped to simplify it somewhat, a considerable amount of bickering went on at pit level over who should receive these allowances and when they should, or should not, be paid.

In line with the proliferation of allowance payments for craftsmen, there was also a substantial increase in the number of craft grades after 1961. The wages structure established in 1955 was very simple, consisting of only three grades – Grade I Plus, Grade I, and Grade II. This remained unchanged up to 1961. In 1961, however, Grade I was divided into two grades: Engineering Craftsmen Grade I, and Craftsmen Grade I. Engineering craftsmen included the jobs of fitters and electricians, but also the related jobs of plater and welder, etc. The other jobs, such as joiner, saddler and painter decorator, remained in Grade I. Craftsmen Grade I Plus were renamed Engineering Craftsmen Grade I Plus. In 1963 a new grade of Engineering Craftsmen Grade II was created to cover Maintenance Men (Self Advancing Chocks). This grade was paid $4\frac{1}{2}$p per shift more than that of Maintenance Men (Mechanical Appliances), which remained in the Craftsmen Grade II. In 1964, new grades (or jobs) of Colliery Electro-Mechanic and Colliery Fitter/Electrician were introduced. Fitters and electricians were removed from Engineering Grade I and paid a rate of $3\frac{1}{2}$p more than the Engineering Grade I rate.

The cumulative effect of these changes was to replace the simple structure established in 1955 with a complex one of seven grades. In some cases, though, the differences in rates between the grades were so small as to be of little real significance.

Ostensibly the proliferation of craft grades, as of allowance payments to craftsmen, was a response to changes in the skill requirements of particular jobs following alterations in apprenticeship rules and new regulations governing mechanics and electricians. In reality it was, like the additional allowance payments to craftsmen, an attempt to meet the pressures created by a shortage of craftsmen by differentiated, rather than general, wage increases. As Lawrence Daly explained to the union's Annual Conference in 1969,[18] 'We are all conscious of the fact that union policy was based on the actual situation in the industry where in certain areas at any rate there were acute shortages of fitters and electricians primarily because of competition from outside industry, and it was arising from that situation that the union agreed to a situation in which there was a higher rate paid for fitters and electricians as compared with other craftsmen in the industry.'

The increased differentiation of the craftsmen's structure probably caused as many problems as it solved. First, it caused resentment among those craftsmen who did not initially gain. This encouraged leapfrogging in the wage structure, as first one class of craftsmen gained a wage advantage and then the remainder argued for similar treatment. It also created an anomalous position between so-called engineering craftsmen (such as fitters and electricians) and the more traditional craftsmen (such as blacksmiths), both of whom had served formal craft apprenticeships. Anomalies also arose between these people and purpose-trained craftsmen, upon whom the industry had become increasingly reliant as the Board sought to ease manpower shortages by fragmenting the more highly skilled crafts into separate jobs. The picture became even more confused, from a wage payment point of view, after the NPLA, which provided for electro-mechanics to be included in power-loading teams. This qualified these craftsmen for the power-loading rate, while others with similar qualifications remained outside teams on the lower standard craft daywage rate. This difficulty was particularly acute during the early years of the NPLA, when NPLA rates varied between coalfields.

These problems combined to convince the NCB and NUM that nothing short of a complete overhaul of the craftsmen's structure would serve to eradicate the difficulties and anomalies which had arisen. A working party was appointed to examine the craftsmen's structure, which reported in 1970, and in March 1971 the NCB and NUM agreed on a reorganised craft-grading structure.[19]

The main feature of the new structure was to provide more meaningful differentiation in terms of craft experience. It also resolved the problem created by the NPLA by taking craftsmen *out* of the power-loading team.

The new structure consisted of four major craft grades, with additional wage differentiation based on the place in the pit at which men worked. The distinction between engineering craftsmen and non-engineering craftsmen was dropped, although one group of engineering craftsmen – colliery electro-mechanics – preserved their own grade. The four grades of craftsmen were: colliery electro-mechanics, craftsmen IA, craftsmen IB, and craftsmen II.

In the new structure, Grade IA was to consist of time-served craftsmen, Grade IB of men who had gained their experience by dint of service, and Grade II of purpose-trained men. In this way, non-engineering time-served craftsmen were put on a par with time-served colliery fitters and electricians whose status was probably exaggerated by the mechanisation programme. Similarly, the previous wage differentials between engineering craftsmen Grade II and other Grade II craftsmen disappeared, both being given the title of Grade II craftsmen. There was still provision for craftsmen who worked on mechanised coal faces to receive the power-loading rate, but they were not to be part of the power-loading team as such, and would henceforth only receive the NPLA rate when on the face. For the remainder of the time they would be on the standard craft rate for their grade. This provision had the advantage of providing greater flexibility in the use of hitherto power-loading team craftsmen, many of whom, although paid a rate superior to that of their craft colleagues, had been under-utilised when in the power-loading team.

The grading structure had the effect of reducing percentage standard shift rates between craftsmen and between craftsmen and non-craftsmen although this trend was reversed again by the 1974 settlement. Taking the period 1955–74 overall, then, there has been some considerable variation in national attitudes towards the relative standing of craftsmen in the industry.

5.5 Underofficials and weekly paid industrial staff

The evolution of the wage structure of underofficials and weekly paid industrial staff up to 1956 was traced in Chapter 2. It will be remembered that, with the exception of East Midlands and Yorkshire, where wages had been slightly higher than in other coalfields, the NCB and NACODS had agreed on uniform weekly wages for deputies and shotfirers and uniform wage ranges for overmen. They had also established a standard wage rate structure for WPIS.

There have been several developments in these wage structures since 1956. The weekly wage of underofficials was calculated on the basis of a six-day week, as opposed to five days for other mineworkers. They received neither overtime rates nor an attendance bonus for the sixth day. This was the source of considerable animosity among underofficials and the Board came under increasing pressure from NACODS to base their wage on a five shift week and pay overtime rates for Saturday and Sunday working.[20]

These pressures, together with the NCB's concern about rising absenteeism among underofficials, resulted in the NCB and NACODS agreeing to introduce a bonus for attendances above a specified number in a given time period. The first bonus was paid in 1957 for attendances in excess of 72 over a thirteen week period. The importance of the bonus in weekly pay since then has increased, as the bonus payable has been raised and the attendances necessary to qualify for the bonus have been reduced. By 1969, the bonus had been fixed at one sixth of the weekly wage. In 1970, the bonus remained at one sixth but was payable on attendances in excess of five per week. In 1971 the bonus was raised to one fifth of the weekly wage.

The differences in standard weekly wage rates which remained in 1956 were gradually reduced as smaller wage increases were negotiated for Yorkshire and the East Midland coalfields than for the other areas. Wage rate equality between deputies and shotfirers was finally achieved in 1969 and for overmen a year later.

While the NCB and NACODS have achieved standard wage rate equality among underofficials, the important differential for these supervisory categories is really the one between them and the people they supervise. In mining, the key differential in this respect has been that between Deputy Grade I and faceworkers. With piecework, the deputy's wage was, more often than not, lower than that of face pieceworkers. It was only with the abolition of piecework that an attempt to raise and standardise the deputy's differential was feasible. Since the NPLA, the NCB has been committed to a 10% differential between Grade I Deputies and NPLA chargemen.

The national weekly wage and grading structure negotiated for WPIS in 1956 was replaced by a job structure in 1963, which incorporated revised conditions of service to bring WPIS into line with those applying to underofficials. Since then, changes in pay and attendance bonus provisions for WPIS, as well as other conditions of service, have tended to follow those of underofficials.

5.6 Summary

The move to uniform daywage rates following the 1955 agreement was by no means as easy to achieve as had been anticipated. Opposition from the highest-paid daywagemen to the standstill on earnings, which would have been necessary to achieve any uniformity in three years, forced the NCB and NUM to raise wage ceilings. Thus, the industry still had men on personal rates in the 1960s. The attempt to cater for abnormal working conditions within the standard wage-rate structure proved unsuccessful and allowance payments were reintroduced. This may also point to the real difficulty of catering for diverse conditions in a standard wage-rate structure.

Although flat-rate wage increases to daywagemen could be justified by the regulatory spin-offs on pieceworkers' earnings and equity considerations, their effect was to compress the daywage structure to a point where differentials became meaningless. 'Preferential treatment' for craftsmen, who received bigger wage increases than daywagemen, reflected their growing importance to the industry as it became more technologically advanced, the general shortage of craftsmen in the economy as a whole, and the craftsmens' advanced status, and power, in the NUM. This provides some evidence of the central negotiators' attempt to accommodate market pressures within the national wage structure. However, the ad hoc approach to craftsmen's wages, enforced by cost considerations, meant that their grading structure became increasingly differentiated and eventually required systematic revision. Our account finally points to the inevitability of repercussive effects within a structure if one part is considered in isolation, and shows that when revising a structure relatively small groups can cause as many problems as those with numerical strength.

PART TWO

NATIONAL WAGE DECISIONS AND WAGE STRUCTURE

6. THE MECHANICS OF PAY DETERMINATION

Thus far we have outlined the major features of national wages policy insofar as it sought to influence wage structure. No attempt has been made as yet to assess the impact of wages policy on actual wage levels and movements. It is to this and related aspects of wage determination in coalmining that we shall turn in this section, which consists of three chapters.

In this and the next chapter we shall consider the extent to which wage movements in coalmining reflected the collective wage decisions of the national bargainers. And in this chapter we shall examine the impact of national wage decisions on wage movements in the industry as a whole. The writer's purpose here will be to attempt a quantitative assessment of the components of aggregate wage movements and the implications of the latter for the internal wage structure of major occupational groups. The impact of wages policy on the regional wage structure is considered separately in Chapter 7, while the implications of the wage trends revealed in Chapters 6 and 7 for the miners' external wage relativity, as well as the latter's more general determinants, will be the subject of Chapter 8.

6.1 National wage bargaining – some popular views

The wage reforms outlined in earlier chapters were one aspect of a national wages policy which had as its overall objective the regulation of miners' wages according to mutually agreed national principles. The centralisation of wage decision making in coalmining, however, has occurred over a period when students of British industrial relations have become increasingly sceptical of industry-wide agreements as a method of regulating pay. This scepticism was most forcibly expressed in the Report of the Donovan Commission,[1] which found Britain to have two systems of industrial relations – a formal system consisting of trade unions and employers' associations on the one hand, and an informal system comprising the actual behaviour of managers, shop stewards and workers, on the other. Donovan concluded that the formal system was frequently of less importance than the informal system in determining pay, and its outmoded existence was the root cause of many of our industrial relations problems. To quote:[2] 'The central defect in British industrial relations is the disorder in factory and workshop relations and pay structures promoted by the conflict between the formal and the informal systems. To remedy this, effective and orderly collective bargaining is required over such issues as the control of incentive schemes, the regulation of hours actually worked, the use of job evaluation, work practices and the linking of changes in pay to changes in performance. In most industries such matters cannot be dealt with effectively by means of industry-wide agreements.'

No one has ever been very clear where coalmining, and the other nationalised industries, fitted into the Donovan analysis of collective bargaining and wage determination, which seemed to favour dismantling established industry-wide agreements and replacing them with a more decentralised system based on the place of work. It could be, given the single ownership of these industries, that Donovan regarded their wage agreements as being of the favoured company type. But there remains the uncomfortable fact that the 'company' agreements of coalmining cover many more workers in establishments probably more dispersed and diverse than those to which most such agreements in the private sector relate. In this case, should not the national agreements of the coalmining industry have been prone to the very same problems which, according to Donovan, industry-wide agreements are ill-suited to cope?

As much of the criticism of industry-wide bargaining's apparent inability to regulate wage levels and movements has been based in part on the observed tendency for indices of actual rates of pay, in sectors where industry-wide bargaining is a feature, to drift upwards faster than indices purporting to measure changes in nationally accepted wage rates, earlier studies of coalmining wages might suggest an affirmative answer to this question. Several writers, for example, have drawn attention to the fact that, up to the mid 1960s at least, actual wage rates in coalmining have tended to rise more quickly than those collectively agreed at national level.

The same writers have offered a number of explanations for this apparent wage drift. In 1956, for example, Professor Turner[3] attributed the divergence during the early years of nationalisation mainly to the scope which the industry's piecework system of the time afforded for local wage revisions in excess of nationally agreed wage awards. This extended his argument of 1952[4] that the dominant position of faceworkers in the NUM was the main factor enabling them to widen an already large wage gap between themselves and datal workers. As Professor Alexander[5] pointed out, however, the high and rising level of coal demand during the first decade of nationalisation was a fairly important background factor to any achievements of pieceworker wage bargaining during that period.

By 1964, however, Turner[6] seems to have adopted the NCB's view that face pieceworkers' earnings tended to follow the pace set by national 'revisions of the daywage structure rather than the reverse. He had previously[7] stressed the importance of such factors as upgrading, changes in the proportion of labour between low- and high-paid jobs, and increased overtime, in explaining a long-run tendency for earnings to rise faster than nationally negotiated wage rates in industry generally, and, although he did not quantify their significance, he thought these factors in a period of rapid change might account for part of the observed wage drift in mining after the mid 1950s. And, over the period 1948–64 as a whole, wage comparability with other sectors was the important stimulus to the continuing wage drift,[8] as 'by one method or another matters have been so arranged that they [miners' wages] have kept pace with wages in outside industry'.

In 1967, Michael Barratt Brown, evidently surprised to find that, in the face of a falling demand for coal in the 1960s, miners' hourly earnings had not continued the relative decline in evidence in the later 1950s, reappraised some of these earlier arguments. While he attributed the apparent maintenance of the wage relativity in part to the national negotiators having to take cognisance of wages outside mining as the industry became increasingly concentrated in the high-wage areas, he also found evidence of local manoeuvring on piecework contracts and overtime which helped to improve miners' pay.[9]

Taken together these writers have presented a very confusing picture of wage movements in mining up to the mid 1960s. Nevertheless, they have all doubted the ability of the national negotiators to regulate wages over the periods they have considered. The central bargainers, of course, went much further in their attempt to regulate wages with the reforms of 1966 and 1971, and piecework – one identified source of wage drift – was eliminated. These reforms undoubtedly made the wage structure more amenable to central direction. Nevertheless, in addition to clarifying what has actually happened to wage movements in mining since nationalisation, we suggest that, whereas it is by no means easy to quantify the contribution of national wage decisions to actual wage movements before 1966, previous attempts to do so have been rather misleading, because on the one hand they have failed to give due attention to the data they have been using, while on the other they have misunderstood the scope and purpose of collective bargaining in the industry.

6.2 Wage drift and central wage control

Wage drift is identified with a tendency for actual rates of pay in the workplace to rise at a faster rate than wage increases negotiated periodically under national wage agreements. The wage drift implies two things: a gap between nationally agreed wage rates and actual wage rates at a moment in time and a tendency for the gap to widen over time.

Wage drift is frequently taken as evidence of the inability of the parties to collective bargaining to regulate wages. It may thus pose problems for a central incomes policy, the influence of which is mainly confined to the institutions of collective bargaining. The task of identifying and measuring the wage drift, however, is complicated by inadequacies in the available data upon which estimates are frequently made, as well as the complexity of the collective bargaining processes to which they may refer.

An accurate estimate of wage drift requires a meaningful statistic of nationally agreed wage rates, as well as of actual earnings to which we can relate it. Most studies of wage drift at an industry-wide level, however, have been based on comparisons of movements in the Department of Employment's (D of E) annual or quarterly indices, called wage rates on the one hand, and earnings on the other. Unfortunately, deficiencies in both series can render them inadequate for the purpose of estimating the impact of national bargaining procedures on actual wage movements.[10]

The D of E's index of wage rates for many industries reflects no more than minimum, or, where relevant, standard wage-rate entitlements detailed in national agreements, arbitration awards or minimum wage-fixing machinery. In some instances it will also make allowance for regional differences in these rates. But the D of E has seldom collected sufficient information about national wage rates or scales above the minimum to take these into account in its wage-rate series. Nor does the series include many other payments which are nationally agreed, such as special allowances for night work, unpleasant conditions, special responsibility, good attendance, merit money, holiday pay, and many more, which nevertheless may be included in the earnings figures. If the growth of any of these items is faster than that of minimum rates, therefore, the D of E base-rates series is likely to under-estimate the contribution of nationally agreed wage items to actual earnings movements.

The earnings series is also fraught with problems. This reflects all the items listed above, together with any items paid at the workplace level. But it is also affected by such things as changes in hours worked and in the relative proportion of workers in different age and sex groups, or engaged in various occupations and wage grades, or paid time and piece rates, the effects of which may increase earnings and produce an earnings drift. Adjustments are frequently made to the earnings series to allow for these problems.[11] But it is likely that such calculations are necessarily somewhat arbitrary, by the very nature of the data available. Nor should we necessarily deduct all the effects of overtime and changing labour force composition when calculating wage drift, for the generous allocation and provisions of overtime, as well as upgrading, may be ways by which the employer seeks to attract and retain labour.

The problems of identifying the wage drift do not end here. For example, even if we could identify all the elements in the wage packet which are the subject of national wage agreements, we would still need to know how these items are affected by national agreements, if we are to measure at all precisely the latter's impact on actual earnings. In the UK early students[12] of the wage drift, and then the D of E, estimated it by subtracting the annual percentage increase in the hourly earnings index adjusted for overtime from the annual percentage increase in the base-rate index. Thus, if the index of hourly earnings adjusted for overtime rose by 11% and the index of hourly rates rose by 10%, the D of E's estimate would give a wage drift of 1%. Some students have also made allowance for the effect of compositional changes in the labour force on hourly earnings before making the wage drift calculations.

While it may be useful for some purposes to compare percentage increases in the hourly rates and earnings series, this method of estimating wage drift assumes that national agreements are *meant* to have a proportionate effect on hourly earnings. But this may not be the intention (although the earnings and basic rates series might actually move in a manner which could *imply* otherwise). It is quite common, for example, for national agreements to affect only the minimum wage rate and the earnings of those workers on it. Again, national wage increases are often specified as flat-rate additions to national rates, which in an industry with widespread local bargaining may only set a floor below which actual wages cannot fall. Thus, the percentage impact of increases in national rates on earnings will depend on the gap which actually exists between actual hourly earnings and nationally agreed hourly rates. Let us suppose, for example, that national wage rates comprise only 50% of hourly earnings and are increased by a flat-rate amount. In this case, the percentage increase in actual hourly earnings arising from the national addition should only be half the percentage increase in national rates. But suppose actual earnings rise by the same percentage as the base rates. On the D of E method of measurement, the wage drift will be zero, despite the intention that earnings should rise by half the percentage increase in rates. The D of E's measurement of drift could easily exaggerate the role of national agreements in actual earnings movements and underplay the wage drift contribution. The conventional method of measuring drift adopted by the D of E is, therefore, no more than an attempt to measure indirectly something which should be measured directly, through a study of collective bargaining arrangements and wage structure.

The above discussion does not exhaust the problems inherent in identifying and measuring the wage drift. Nevertheless, it does illustrate that conclusions about the effectiveness of national collective bargaining procedures in regulating earnings based solely on crude comparisons of the wage series described above could prove highly misleading. This is particularly true of coalmining for much of the period covered in this study.

6.3 THE MECHANICS OF WAGE DETERMINATION

In the light of the foregoing discussion we shall preface this analysis of wage movements in coalmining by summarising the various ways in which miners' pay could have been raised since nationalisation. This outline, which relies heavily on our account of wage structure in earlier chapters, will help us place national wage decisions in their proper context when we come shortly to analyse their contribution to actual wage movements. It will also highlight many of the problems inherent in the wage drift concept when applied to coalmining.

6.3.1 *National minimum, standard wage rates and flat rates*

Let us start by listing those wage items which can be attributed directly to national wage negotiations. The most important for some time has been the *national minimum rate*, which has underpinned miners' wages since 1942 and been increased at frequent – usually yearly – intervals. As the minimum wage has been taken by most students of the wage drift in coalmining to represent increases in nationally agreed base rates, its role in wage determination requires careful analysis.

There was a substantial number of men (mainly daywagemen) whose wages had to be 'made-up' to the minimum wage when it was first introduced. Wages above the minimum, but below agreed wage ceilings, were only to be increased by a proportion of the monetary amount by which the minimum was raised. Those miners receiving wage rates above the ceilings received nothing from minimum wage-rate increases – pieceworkers formed the bulk of this latter group. They, and other workers whose wages were above the agreed ceilings, only benefited from national increases insofar as certain elements in their wage packet (such as overtime payments, waiting-time pay and make-up for working in abnormal conditions) were geared to changes in the minimum rate. The minimum wage has always been slightly higher for underground workers than for surface workers – the differential being justified by the more arduous physical conditions associated with underground work.

With the new daywage structure in 1955, it became strictly incorrect to speak of minimum base rates in the case of daywagemen. The rates laid down in the 1955 agreement, and subsequent revisions of that agreement, were standard in the sense that daywagemen were entitled to no more, and no less, than the rate specified in the national agreement. In fact, though, the lowest daywage rate at the surface and underground also represents the lowest minimum rate payable to adult mineworkers.

With the exception of craft rates, standard daywage rates were usually raised by the same *monetary* amount in national wage negotiations after 1955. But those daywagemen on personal rates above the standard rate after 1955 only received a proportion of the monetary increases up to an agreed ceiling which was revised from time to time. Clearly, then, the percentage change in the lowest standard rate alone will overestimate the overall percentage increase in day rates implied by national wage rate revisions.

Before the national wage structure revisions in 1966 (in the case of face pieceworkers) and 1971 (in the case of pieceworkers working behind the coalface), pieceworkers had only benefited directly from one 'national' wage award since nationalisation. That was in 1954, when, in return for a standstill on piece rates, pieceworkers received a flat-rate payment of 2/11d per shift. *Other flat rates* in the workers' pay packet at nationalisation (such as the 'skilled shilling' and the 1940 War Additions to wages payment) remained unaltered after nationalisation. In 1955, all these payments were consolidated into the standard rates applicable to daywagemen, although they continued to be paid to pieceworkers.

6.3.2 *Nationally agreed variable supplements*

In addition to the above items, whose size is determined absolutely by national wage agreements, several items have figured in the miners' wage packet which are payable under national wage agreements but whose actual size has not been fixed by them *absolutely*.

It was agreed in 1947, for example, that a bonus of 20% of normal weekly earnings, in the case of daywagemen, and of 16% in the case of pieceworkers, would be payable for the shift no longer worked under the terms of the Five-Day-Week Agreement. In the case of daywagemen on the minimum rate, and those on the standard rate since 1955, the amount of so-called *Five-Day-Week Bonus* (FDWB) payable was determined wholly by national agreement. In the case of daywagemen on rates above the minimum prior to 1955, and pieceworkers throughout the post-war period, however, the monetary value of the bonus depended on the locally agreed wage rate, and, for pieceworkers, on the level of normal earnings, which would vary with effort and output.

The conditions attached to the payment of the bonus changed quite substantially between 1947 and 1972 when, after the national strike of that year, it was finally agreed that it should be consolidated into the basic shift rate. When it was first introduced the bonus was only payable if *all* five normal shifts were worked. After 1952, however, a man absent from work for medical reasons was entitled to a part of the bonus calculated on the time actually worked, and in 1957 that provision was extended to voluntary absence as well.[13] It was still the case, however, that a man who was recorded as having been on strike during a particular week lost the entire bonus for the week in which the strike occurred.

Even after the daywage structure was introduced in 1955, the NCB and NUM agreed, as we have already noted in Chapter 5, to a series of *nationally agreed allowance payments* for such things as working in abnormal conditions and for performing extra duties, like carrying chemical buckets and first-aid kits. These have all been increased from time to time. Allowances were also agreed nationally for craftsmen in recognition of their differing skills and qualifications.

By an agreement of June 1966, daywagemen working at night were entitled to a *night allowance* based on the starting time of their shift. The allowances were not subject to overtime and week-end premia and were excluded from earnings when calculating the FDWB.

Since 1947 men working in the coalmining industry have received *payments for statutory holidays* and, since 1954, payment has been extended to *annual holidays*. In 1964 it was agreed further that, rather than shorten the length of the working week, miners should take seven rest days of their own choice without loss of pay. In 1958 the Board introduced a supplementary *sickness pay* scheme for its manual workers, the benefits from which vary according to age.

Because the average earnings statistics published by the NCB and sent to the Department of Employment are calculated by dividing the total wages actually paid, including those for holidays, rest days and sickness, by the shifts or weeks *actually* worked, these supplementary payments are included in the published earnings figures of miners.

The important thing about all these supplementary payments, however, is that, although they are all payable under national wage agreements, the amount payable varies according to factors such as the age of the worker, his health, attendance, strike record, working conditions, etc. But as we shall see shortly, any estimate of the impact of national agreements on actual wage levels and movements which ignores these items is likely to grossly underestimate the contribution of national wage decisions to actual wage levels and movements.

6.3.3 *Hours of work*

Overtime is another wage element the rate for which is specified in national agreements. The actual hours worked at overtime rates, though, are a matter for local discretion. Overtime may be worked to compensate for labour shortages – under the Extension of Hours Agreement of 1948, which lasted until 1956, miners were expected to work overtime hours as necessary. Again, many jobs in the pit (such as maintenance) can only be performed at weekends or such times as the machinery has stopped. Alternatively, overtime could be agreed upon simply as a means of boosting earnings. Some pits have overtime rotas so that overtime can be spread evenly among the men who want it. In other pits there are no such agreements. Overtime premium rates have changed since nationalisation, but are currently time-and-a-half for mid-week overtime and double time at weekends. Overtime is *not* conditional upon a man having worked all five normal shifts.

6.3.4 *Grading and compositional changes in the labour force*

The industry's wage bill, and thus the average wage, will also have been affected by nationally agreed changes in job *grading* necessitated by technological change, as well as by the addition of new grades at the top of the daywage scale, and the craftsmen's structure.

The average wage may also have been affected by changes in the proportion of labour employed between the coalfields, given the national wage disparities between regions and pits. It may also have been affected by changes in the distribution of labour between places of work in the pits and in its age distribution. Whereas the adult wage was payable after 21 years of age, it has been payable at 18 years of age since 1974. A miner tends to rise to an earnings peak in his early forties and after that takes on less onerous and lower-paid work.[14] As the labour force has aged quite considerably, this will have had a negative impact on earnings movements. To the extent that all these factors may be predicted, however, they could be allowed for by the national negotiators.

6.3.5 *District and pit agreements*

Even if one were able to estimate at all accurately the contribution of the above factors to actual wage levels and movements, the resultant figure would still be an inadequate measure of the role of formal collective bargaining as provided for by national agreements in wage determination in coalmining. In the absence of national bargaining procedures before 1942, miners' wages were dependent on pit and district agreements. And after nationalisation *collective bargaining at pit and wage district level* was explicitly provided for by national agreement. Indeed, one could argue that coalmining has had, not one formal system of industrial relations, but three, for much of the post-war period. There are good reasons, therefore, why an index of nationally agreed wage rates in coalmining should include those wage items agreed *formally* through the pit and district conciliation procedures.

Prior to the NPLA and the Third Daywage Structure, the main items for negotiation through the pit conciliation scheme were *piece-workers' price lists.* Many of the early power-loading agreements were also negotiated at pit level. By the mid 1950s, however, most power-loading rates were, with the exception of Lancashire and Kent, covered by district agreements which also had to be ratified at national headquarters.

Wages could be increased by revising price lists and district agreements. But as all these agreements, except the power-loading agreements in Scotland and Durham, were of a piecerate type, wages would rise with increases in productivity.

But whereas pieceworkers' wages could rise for legitimate reasons provided for by the collective bargaining machinery, they could also rise for reasons which could not be so described, many of which were outlined in Chapter 4. For example, although many changes in piece rates were collectively agreed, physical conditions in the pit would change so fast that new rates were often agreed on the spot, rather than through formal procedures. Again *allowance payments* were frequently agreed to cover abnormal conditions. But the legitimacy of these payments would be difficult to establish. Another practice was to pay rates above those collectively agreed for a special task, or one which was not covered by a job identified in the grading structure. But, without conducting a very thorough analysis of pit price lists and the conditions which applied in all the pits after nationalisation, it would be impossible to ascertain with any accuracy how much of the locally agreed element in pay since nationalisation fell outside the process of formal collective bargaining.

Following the NPLA in 1966 and the Third Daywage Structure in 1971, pieceworkers and taskworkers were gradually absorbed into the national standard rate structure. As the wage rates under these agreements were based on crude averages of existing earnings levels, however, they could be regarded as an ex-post legitimisation of previous malpractices.

Of the other wage elements negotiated at district level, the most important are various *payments made in kind*. These consist of free coal and subsidised housing. Their value is assessed at the market price of the commodity in the local area, and thus varies quite substantially between the coalfields, and over time, depending on market factors.

6.4 The composition of pay

It would clearly be of great advantage, when analysing the impact of collective bargaining on actual pay levels and movements, if we could isolate and quantify each of the pay elements listed in the foregoing section. Although we are unable to do this as thoroughly as we would wish, we can, with the unpublished financial returns of the NCB, go much further in this respect than has been possible hitherto on the basis of published data.

The financial returns of the NCB give, on an annual basis, the allocation of the total wage bill between items payable under various national agreements to miners in each of the major wage-payment categories (daywage and piecework) at each place of work (the pit surface, coalface and elsewhere underground), as well as the number of shifts worked in normal and overtime hours. From these data the writer has been able to compute directly those parts of the annual earnings per manshift for the whole industry, and each wage-payment sub-group within it, attributable to overtime working, the five-day-week bonus, nationally agreed flat-rate additions to wages, holidays and rest day agreements, sickness pay and allowances in kind.[15] The data for shifts worked at each place of work by men in each payment category also enabled him to calculate, from year to year, the approximate effect of changes in the proportion of shifts worked by these groups on total EMS. This was done by holding wages constant and allowing the shift weights to vary. This we call the 'payment systems and grading effect'. From data for the individual coalfields, and by the same method, we can isolate these effects on earnings from those due to changes in the proportion of shifts worked in the various coalfields. This we call the 'regional employment effect'.

When we sum all the above items and subtract them from the total EMS we are left with a figure which represents the contribution to EMS of the nationally agreed minimum and standard wage rates, special nationally agreed allowance payments, other than those already referred to, and district and local payments above the nationally agreed rates and local allowances. We are unable to subdivide this figure into its constituent parts with any precision. Nevertheless, we can calculate a figure for the weighted average of the surface and underground national minimum wage,[16] which we can subtract from the remaining earnings figure. Having done this we are left with a 'residual' wage item, which will include national payments above the minimum wage as well as district and workplace supplements. We can calculate this residual separately for pieceworkers and daywagemen and, knowing the proportion of shifts worked by each group, can estimate the (weighted) contribution of each group to the total residual in EMS.

Table 6.1 shows the actual and percentage contribution of the wage items listed above to the annual average EMS of the industry as a whole in various sub-periods between 1948 and 1973. One of the most interesting features of the data is that, while the relative importance of the minimum wage rate in actual pay was falling over the post-war period, this was largely offset by the growth of other wage items payable under national agreements. The combined value of items 3–7 in the table, for example, increased by 350% over the period 1948/52–1968/73, while the minimum wage rose by just over 200%. In the period 1968–73, 30% of the miner's average shift rate came from items payable under national agreements other than the basic shift rate, as compared with 24% in the period 1948–52. The largest single item under that heading was traditionally the five-day-week bonus. Before its consolidation into national shift rates in 1972, the value of the FDWB was growing faster than the minimum wage. This was mainly because of the nationally agreed relaxation of the conditions under which it was payable. But the fastest growth in

Table 6.1. *Make-up of average earnings per manshift of all mineworkers 1948–73 (shillings; percentages in brackets)*

Wage item	1948–52	1953–57	1958–62 (Annual averages)	1963–67	1968–73
1. Total EMS	37.0	52.0	66.8	85.4	132.6
2. National minimum rate	22.6	29.7	35.9	43.5	70.6
	(61.2)	(57.1)	(53.7)	(50.9)	(53.1)
Difference (1–2)	14.4	22.3	30.9	41.9	62.0
Other items payable under national agreements					
3. Overtime premium	2.1	3.3	2.3	3.8	6.1
	(5.7)	(6.3)	(3.4)	(4.4)	(4.6)
4. FDWB	3.8	5.1	7.3	8.8	9.0
	(10.3)	(9.8)	(10.9)	(10.3)	(6.8)
5. Holiday pay	1.5	2.8	3.8	6.5	12.8
	(4.1)	(5.4)	(5.7)	(7.6)	(9.6)
6. Sickness pay	—	0.2	0.4	0.7	2.6
		(0.4)	(0.6)	(0.8)	(2.0)
7. Allowances in kind	1.5	2.3	3.8	5.3	9.1
	(4.1)	(4.4)	(5.7)	(6.2)	(6.8)
Σ 3–7	8.9	13.7	17.6	25.1	39.6
	(24.2)	(26.3)	(26.3)	(29.3)	(29.8)
'Compositional' effects					
8. Regional employment changes	—	—	0.1	0.4	0.8
			(0.1)	(0.5)	(0.6)
9. Payment system and grading changes	—	0.1	1.6	1.7	2.0
		(0.2)	(2.4)	(2.0)	(1.5)
10. *Residual* (1 – Σ (2–9))	5.3	8.5	11.6	14.8	19.9
	(14.6)	(16.3)	(17.3)	(17.3)	(15.0)
Weighted contribution of:					
(*a*) pieceworkers	5.1	7.3	9.9	11.7[a]	12.1[a]
	(13.8)	(14.0)	(14.8)	(13.7)	(9.1)
(*b*) daywagemen	0.3	1.2	1.7	3.1	7.8
	(0.8)	(2.3)	(2.5)	(3.6)	(5.9)

[a] Includes nationally agreed shift rates payable under the National Power Loading Agreement and Third Daywage Structure.

Source: Calculated from the Financial Returns of the NCB.

value terms has been in holiday pay, which was nearly 10% of recorded EMS in the period 1968–73, and in 1973 itself was 13% of EMS. The jump in the value of allowances in kind in the period 1968–73 reflects the general rise in the price of coal and rents.

As the monetary value of the overtime premium is related to basic rates, it is not surprising that it should have increased over the post-nationalisation period. It declined in the late 1950s and early 1960s, due to the abandonment of the Extension of Hours Agreement, and in part reflects the demand for mining labour. It is also interesting to note that the overtime premium value has otherwise retained a fairly constant relativity to the minimum wage rate. While this does not rule out collusion on overtime working, to ensure pay above the level it would otherwise have been (one might have expected overtime in the mid 1960s to have stayed well below its early 1950s level), it does suggest that national agreements themselves have been a major factor in the growth of the overtime premium.

The estimate of the effect of changes in the proportion of the labour force, between coalfields on the one hand, and between payment methods on the other, suggests that in the period 1968–73 EMS was 2.8 shillings above what it would have been had the distribution of labour remained constant between payment methods and coalfields after 1948. The most important effect was that due to changing labour proportions between wage payment categories at various places of work in the pit, and most of that effect was confined to the late 1950s when the industry was going through a major reclassification of its labour force and applying the 1955 daywage grading system. However, there was a steady increase in the proportion of pieceworkers in the industry between 1955 and 1966, particularly in elsewhere underground jobs, which, we believe, did not wholly reflect national decisions. We shall return to this later.

Let us now examine the residual wages item. It must be emphasised that, because this necessarily includes some nationally agreed pay elements above the minimum wage, it is not a wholly accurate measure of district and workplace wage supplements in pay. The major anomaly in this respect is after 1966, when the residual for pieceworkers includes the national power-loading rates and (after 1971) the consolidated rates of the Third Daywage Structure. Again, as our data refer to all mineworkers, they include the wages of juveniles as well as certain foremen grades, such as weekly paid industrial staff and underofficials. Unfortunately the data do not enable us to distinguish these groups from other daywagemen in our table over the whole period. Their inclusion in the daywage EMS figure helps to bring the shift rate net of national additions closer to the weighted minimum wage. But with these reservations, and the exception of pieceworkers after 1966, the residual probably gives a not too distorted picture of the *relative* importance of wage items other than those specifically paid under national agreement for most of the post-nationalisation period. The residual item has grown in each period for which data are given. By far the largest part, though, was not unexpectedly associated with the piecework structure. And whereas the contribution of the pieceworker residual to total EMS grew by 6.6 shillings over the period 1948–52 to 1963–67 (when it most accurately reflects local and district bargaining), the contribution of the daywage residual to total EMS increased by only 2.8 shillings a shift over the same period. In percentage terms the contribution of local wage additions to total EMS probably grew from just over 5% in the period 1948–52 to 14% in the period 1963–67. Of the latter, nearly 75% was due to local piecework additions to pay. We shall examine the residual item in more detail shortly.

6.5 National wage decisions and wage movements

While the above analysis provides an interesting insight into the composition of aggregate EMS in the industry as a whole, it is only a preliminary step in analysing the mechanics of wage *movements*. For example, although the minimum wage acted as a floor below which actual EMS could not fall, it would be incorrect (for the reasons already explained) to assume that EMS should have risen by the full monetary increase in the minimum wage, particularly in the case of pieceworkers. Before we attempt to take account of the intentions of the national bargainers, however, we ought to point to one interesting implication of our analysis so far for previous attempts to estimate wage drift in coalmining.

All the past studies of wage drift in mining[17] have estimated it by comparing percentage changes in the minimum shift rate and actual EMS, net of allowances in kind. Even if we ignore the validity of the assumptions about the intended impact of the minimum wage increases on actual EMS underlying that calculation, it is immediately apparent that, to the extent that such studies have completely ignored items in EMS paid under national agreements other than the minimum wage, they will have inadvertently given a misleading impression of the role of wage drift and its source.

The figures given in Table 6.2, for example, show that, employing the Department of Employment's estimating procedures, an earnings drift of 1.4% per annum over the period 1948–66 reduces to 0.6% per annum if one compares percentage changes in the EMS with changes in the combined statistic of minimum wage rates, five-day-week bonus and holiday and sickness pay, instead of that for minimum wage rates alone. Similarly, a wage drift estimate of 1.3% based on the minimum wage comparison reduces to 0.3% if the composite figure of national wage additions is used instead.

Although our estimate of wage drift based on the minimum wage rate and EMS comparisons is 0.3% higher than the 1% per annum estimated by Professor Turner,[18] it is interesting to note that a substantial part of the 'wage drift' he identified, using minimum wage rates and Department of Employment methodology, could be accounted for by nationally agreed wage elements in the wage packet. Furthermore, our statistical comparisons would suggest that by the period 1963–66 the industry was experiencing a negative wage drift of −0.5% per annum, as compared with a positive drift of 0.6% per annum when EMS changes are compared with those of the minimum wage alone. While both estimates lead one to the view that national wage decisions in coalmining were becoming increasingly effective in regulating wage movements after the mid 1950s, crude statistical comparisons such as these, which pay scant attention to the intentions of the national negotiators, are rather misleading if one wishes to

Table 6.2. *Some estimates of earnings and wages drift based on Department of Employment procedures*[a]

	Estimate A		Estimate B	
	Earnings drift %	Wage drift % (annual averages)	Earnings drift %	Wage drift %
1948–51	2.1	1.6	1.3	0.8
1951–54	1.6	1.7	0.6	0.7
1954–57	1.6	1.1	0.7	0.2
1957–60	1.3	1.9	0.1	0.5
1960–63	1.1	0.8	0.6	0.2
1963–66	0.7	0.6	0.3	−0.5
Average	1.4	1.3	0.6	0.3

[a]For the purposes of calculating *wage drift* we have excluded the value of regional and skill effects and the overtime premium from actual EMS which, to achieve consistency with certain estimates, also excludes allowances in kind. In *estimate A* we have compared annual percentage changes in the resultant EMS figure with those of the national minimum wage rate, expressed as a weighted average of the underground and surface minimum rate. In *estimate B* we have compared percentage changes in EMS with the percentage change in the sum of the minimum wage, five-day-week bonus, and holiday and sickness pay. This has not been extended beyond 1966, as changes in wage structure after that year make such calculations rather meaningless and our purpose here is only to illustrate the misleading impressions created by *estimate A*.

understand the actual mechanics of pay movements. Let us turn then to examine this problem in more detail.

At the risk of becoming repetitive, it is important to appreciate that before 1955 national wage increases were applied only to the minimum wage. Daywagemen above the minimum were intended to receive *less* than proportionate increases in their base rates. In 1951 approximately 60% of daywagemen were on the minimum wage rate.[19] If we now assume that the remaining daywagemen were evenly spread across the wage bands up to the agreed ceilings, we would estimate that only 45% of the minimum *increase* was actually reflected in the EMS of these miners. Thus, before 1955, we would estimate that daywage base rates should have risen by not more than 82% of the money increase in the minimum wage. Pieceworkers were to receive nothing from these increases unless they fell into an appropriately low wage-rate band: their wages were left to local negotiations. As pieceworkers worked some 36% of shifts in the industry in the 1950s,[20] it is doubtful if more than 53% of any increase in daywage shift rates should have been transmitted to actual EMS.

We must now raise this figure slightly to allow for the fact that a few pieceworkers were affected by adjustments in the minimum insofar as it affected fall-back rates when equipment was not working. The effect of increases in the minimum on their overtime payments and, where appropriate, five-day-week bonus is already catered for in the relevant statistics. We think it unlikely that pieceworkers' wages otherwise benefited directly by more than 5% of any increase in the minimum wage. Thus, in view of the proportion of total shifts worked by pieceworkers, the direct effect of a minimum wage adjustment on the all industry EMS figure was about 1.8% in their case. In total, then, we doubt if more than approximately 55% of any increase in the minimum rate should have been transmitted to the average EMS of the industry as a whole in 1950. This proportion is likely to have increased slightly between 1948 and 1954 as wage increases brought the wages of an increasing proportion of workers into the range which qualified for the full amount of minimum additions. We have no precise way of knowing what adjustment should be made for this, but if the spread of earnings just before the 1955 daywage revision is an indication, an adjustment of −1% a year before 1950 and of +1% a year thereafter would seem appropriate. Thus our weights will range from 52% in 1948 to 58% in 1954.

We know that immediately after the daywage revision some 80% of daywagemen were on standard rates which were raised by the same monetary value as the minimum wage, and that 18% of the remainder qualified for less than proportionate increases. On the

Table 6.3. *The percentage point contribution of estimatable wage items to the annual percentage change in average earnings per manshift for all mineworkers 1948–72*

Period	EMS % change	% point contribution to EMS: Minimum wage	Five day week bonus	Holiday pay	Sickness pay	Allowances in kind	Overtime premium	'Compositional' effects	*Residual*: Piece-workers (P)	Daywage (D)	P & D
48/49	3.6	—	0.6	0.6	—	0.3	0.3	—	1.8	—	1.8
49/50	3.5	0.20	—	—	—	0.2	0.5	—	1.9	0.6	2.5
50/51	9.3	3.00	0.3	0.8	—	0.6	1.4	—	3.9	−1.0	2.9
Average	5.47	1.07	0.30	0.47	—	0.37	0.73	—	2.53	−0.1	2.43
51/52	11.3	2.90	0.80	1.80	0.20	0.40	0.20	—	3.40	1.70	5.10
52/53	5.1	1.18	0.46	0.46	0.22	0.22	0.46	—	1.50	0.55	2.10
53/54	5.3	1.90	0.66	—	—	0.44	0.44	—	1.60	0.30	1.90
Average	7.23	1.99	0.64	0.75	0.14	0.35	0.37	—	2.17	0.85	3.02
54/55	6.48	1.90	0.63	0.42	—	0.21	0.63	−0.21	2.09	0.84	2.93
55/56	9.43	3.43	0.98	0.20	—	0.98	0.39	−0.39	2.94	0.79	3.73
56/57	7.72	1.76	1.08	0.90	0.18	0.18	0.36	1.80	2.33	−0.90	1.43
Average	7.88	2.36	0.90	0.51	0.06	0.46	0.46	0.40	2.45	0.24	2.69
Average 1948/57	6.85	1.80	0.71	0.58	0.06	0.39	0.52	0.13	2.38	0.32	2.70
57/58	4.00	0.82	1.17	0.66	0.17	0.83	−2.00	1.66	1.42	−0.75	0.67

58/59	2.40	0.90	0.64	—	—	0.32	−1.12	−0.48	0.72	1.36	2.08
59/60	2.81	0.34	0.16	0.16	—	0.31	0.47	0.31	0.78	0.31	1.09
Average	3.07	0.69	0.66	0.27	0.06	0.49	−0.88	0.50	0.99	0.31	1.28
60/61	6.09	2.13	0.46	0.15	0.15	0.16	0.61	0.15	1.22	0.61	1.83
61/62	3.87	0.90	0.14	0.14	—	0.43	0.17	—	1.08	1.08	2.16
62/63	6.08	1.74	0.69	0.97	—	0.41	0.69	0.28	1.52	−0.28	1.24
Average	5.35	1.59	0.43	0.42	0.05	0.33	0.48	0.14	1.27	0.47	1.74
63/64	5.34	1.09	0.39	1.56	0.13	0.26	0.26	—	1.17	0.52	1.69
64/65	5.93	2.22	—	1.24	0.12	0.25	0.87	—	1.24	—	1.24
65/66	5.13	1.75	0.70	0.47	—	0.70	0.23	—	1.17	0.17	1.34
Average	5.47	1.69	0.36	1.09	0.08	0.40	0.45	—	1.19	0.23	1.42
Average 1957/66	4.62	1.32	0.48	0.59	0.06	0.46	0.05	0.21	1.14	0.33	1.47
66/67	3.88	0.89	0.55	0.22	0.11	0.22	0.33	0.22	0.33	1.00	1.33
67/68	4.91	1.72	0.53	0.53	0.11	0.21	0.43	—	0.43	0.96	1.39
68/69	6.01	2.42	0.61	0.61	0.10	0.51	0.51	0.10	1.06	—	1.06
Average	4.93	1.68	0.56	0.45	0.11	0.31	0.42	0.11	0.61	0.65	1.26
69/70	11.53	2.82	1.25	0.38	1.93	0.86	0.67	0.10	1.92	1.54	3.46
70/71	14.64	6.80	1.98	2.07	0.17	0.09	−0.34	0.09	3.27	0.52	3.79
71/72	18.41	12.40	−7.06	2.33	0.08	2.03	1.05	—	3.76	3.83	7.59
Average	14.86	7.34	−1.28	1.59	0.72	0.99	0.46	0.06	2.98	1.96	4.94

Source: Calculated from NCB, Financial Returns, 1948–72.

same assumptions as before, therefore, we would estimate that 95% of the national wage increase would be transmitted to daywagemens' EMS but only 65% to the EMS of the industry as a whole. The ceilings were progressively raised after 1957, so we shall increase our weight to 70% in that year and keep it constant thereafter.

On the basis of these probably realistic estimates of how national base-rate adjustments should have affected EMS, we are now able to compute directly the percentage point contribution of various estimated wage elements to the annual percentage change in the all industry EMS. The results of this exercise are given in Table 6.3. Under the heading 'compositional effects' we have combined the contributions of the 'regional employment effect' and the 'wage payment systems effect'. The contribution of the 'residual' wage item is again divided between pieceworkers and daywagemen. In the case of pieceworkers, this item represents the *weighted* contribution of pieceworkers' wage increases *net* of all the national wage agreement elements previously outlined. Before 1966 it reflects local and district wage adjustments; after 1966, however, it includes wage increases given under *national* wage agreements, namely, the NPLA and the Third Daywage Structure Reform. We shall examine shortly whether the national wage increases for pieceworkers after 1966 correspond with the intention of these agreements. In the case of daywagemen, the residual items will also include some increases above standard rates provided for under national agreement which we cannot identify separately. These were very small prior to the late 1960s, however, so although the residual has to be treated with some caution, the table probably gives a fairly accurate breakdown of annual wage increases according to their level of negotiation for most of the period covered. We shall, in any case, delve deeper into the significance of the residual shortly. Table 6.3 is supplemented by Table 6.4, which expresses the percentage point change in each estimated wage item as a percentage of the total change in EMS.

Our calculations in the two tables reveal that, while an increasing proportion of the annual change in the EMS stems from increases in the minimum wage over the period studied, the proportion of increases from this source only one exceeded 50%. Indeed, for much of the post-war period increases in the minimum wage accounted directly for less than 30% of the overall increase in EMS, and during the period 1948–51 – when for most of the time the government was operating a 'wage freeze' – no more than 20% of the total increase in EMS on average. Over the period 1948–57 as a whole, increases in the minimum wage alone would have increased EMS by only 1.8% per annum, and over the period 1957–66 by

Table 6.4. *The percentage shares of estimated wage items in percentage changes in earnings per manshift 1948–73*

	Periods							
	1948–51	1951–54	1954–57	1957–60	1960–63	1963–66	1966–69	1969–73
Percentage of EMS change contributed by changes in:	Percentage shares (annual averages)							
1. Minimum wage and standard day rates	20	28	30	22	30	31	34	49
2. FDWB	5	9	11	21	8	7	11	−9
3. Holiday pay	9	10	6	9	8	20	9	11
4. Sick pay	—	2	1	2	1	1	2	5
5. Allowances in kind	7	5	6	16	9	7	6	7
Total 1–5	41	54	54	70	56	68	62	63
6. Overtime premium	13	5	6	−29	9	8	9	3
7. 'Compositional' effects	—	—	5	16	3	—	2	—
8. 'Residual'	44	42	34	42	33	26	26	33
Contribution of:								
(*a*) Pieceworkers	46	30	31	32	24	22	12[a]	20[a]
(*b*) Daywagemen	−2	12	3	10	9	4	13	13
Total EMS*	100	100	100	100	100	100	100	100

[a]Includes increases under the NPLA and Third Daywage Structure.
Notes: The columns may not sum exactly to 100 due to rounding effects.
Source: Table 6.3 (with modifications as described in text).

1.3% per annum. However, if we now add that part of increases in EMS arising from the five-day-week bonus, holiday and sickness pay, as well as allowances in kind, to that emanating from minimum wage adjustments, the proportionate impact of items payable under national agreements rises quite considerably – from 41% over the period 1948–51 to over 60% in the late 1960s and early 1970s. In the period 1957–60, an abnormally high proportion of the increase in EMS was attributable to these items (70%) because of the sharp cut-back in overtime premium payments. The increases in nationally agreed items other than the minimum wage raised EMS by 1.7% per annum over the period 1948–57. It is interesting to find that during the wage freeze of 1948–50 EMS would have risen by 1% per annum because of increases in these national supplements. Over the period 1957–66, these national supplements raised EMS by 1.6% per annum and were a more important source of EMS increase than the minimum wage.

While these nationally agreed supplements to pay were clearly an important item in aggregate wage movements and in miners' earnings, percentage point increases in their contribution to EMS were negatively related to the percentage point contribution of the minimum wage ($r = -0.52$). This suggests, not surprisingly, that the NUM had to make a decision between increases in any one of these items and increases in the other. This gives added legitimacy to their inclusion in an index of nationally agreed wage rates. Furthermore, it means that the behaviour of these nationally agreed supplements is, in part, an explanation of movements in basic wage rates, and vice versa.

Overtime premium pay has made a very volatile contribution to miners' pay increases: over the period 1948–57, when the industry was operating the national Extension of Hours Agreement, it raised EMS by some 0.5% per annum and constituted some 13% of the annual increase in EMS. In the period 1957–60, when the industry went into its post-war decline, and working hours were severely curtailed, the overtime premium fell quite substantially and had a negative impact on EMS increases of approximately −1.0% per annum. Thereafter it raised EMS by some 0.5% per annum. The depressing effect of the cut-back in overtime on EMS in the period 1957–60 was partly offset, however, by a positive weighting effect in EMS resulting from changes in the labour force composition. The most important compositional effect was that due to an increase in the proportion of miners on piecework-payment systems and an increased proportion working elsewhere underground, rather than that due to changes in the proportionate distribution of the labour force between the various coalfields. As most of the increase in EMS from the former source was attributable to the NCB's labour regrading exercise as between face and elsewhere underground tasks in the late 1950s, however, it may be regarded as the product of national decisions. After then it can be explained by a steady increase in the proportion of pieceworkers in elsewhere underground jobs which may be associated, in part, with technical change.

Let us now examine the role in EMS changes of the residual wage item, which we take as a proxy for local and district bargaining. The contribution of items under this heading tended to fall until the period 1969–72, when it undoubtedly included some of the effect of national changes in the wage structure. In the period 1948–51, wage items under this heading raised EMS by 2.4% per annum and accounted for some 44% of the annual average increase in EMS. Between 1963 and 1966 they raised EMS by 1.4% per annum and accounted for about 26% of the total increase in EMS. During the first decade of nationalisation (1948–57) items agreed in local bargaining raised EMS by 2.7% per annum, which represented 36% of an increase in EMS of 6.9% per annum. Over the period 1957–66 the contribution of these locally agreed items to increases in EMS fell to 1.5% per annum which represented 33% of a lower increase in EMS of 4.6% per annum.

Not surprisingly, most of the EMS increase from local bargaining was associated with the piecework structure. At most, total EMS during the period 1948–66 was raised by 0.2% per annum by local increases to daywagemen. Local and district wage increases to pieceworkers served to raise average EMS by approximately 2.4% per annum during the first decade of nationalisation. In the second decade, however, wage increases to pieceworkers at local level raised overall EMS by approximately 1.1% per annum – a decrease which was appreciably larger than that in the contribution of the minimum wage to percentage changes in EMS. In view of the fact that the proportion of shifts worked by pieceworkers increased from 36% in 1956 to 44% in 1966, movements in the pieceworker residual item imply a much tighter control over piece rates in the second period of nationalisation than in the first. We shall show shortly that the contribution of the so-called pieceworkers' residual to the annual increase in EMS after 1966 accords very well with what might be expected from national increases in the NPLA rates and increases in wage rates given in the course of the Third Daywage Structure reform completed in 1971.

The negative residuals for daywagemen in the years 1956–58 imply a reduction in the gap between actual and standard wage rates and say much for the effectiveness of the ceiling's policy in bringing daywagemen on personal rates within the standard wage rate structure. The fact that the daywage residual is then positive in every year except 1962–63 confirms our earlier impression that the ceilings policy became increasingly less effective thereafter and that pit managements tended to pass on the full increase in the minimum rate to personal rate and 'mark-time' men.

We have noted already, however, that our residual wage item picks up some nationally agreed increases in pay over and above the minimum wage, which cannot be separately identified in the data at our disposal, particularly after 1966. For example, the fairly large residual for daywagemen in the period 1951–52 is partly explained by the revision of the craftsmens' wage structure in 1952, whereas the residual in 1961–62 is undoubtedly enhanced by the addition of new wage scales for craftsmen who also benefited from slightly larger wage increases than the daywagemen in 1962. Again the residual in the period 1954–56 almost certainly includes some of the increases over and above that of the minimum wage, following the revision of the daywage structure in 1955. The daywage residual also includes increases in the numerous allowance payments negotiated at national level for such such things as carrying chemical buckets, working in dust and, after 1966, special night shift payments. When all these considerations are taken into account, the contribution of the residual wage item for daywagemen to total EMS changes could well have been as little as 0.2% per annum.

6.6 Some implications for internal wage differentials

While the previous analysis has enabled us to see the role of local bargaining in aggregate wage movements in its correct perspective, we must necessarily examine its impact on internal wage differentials if we are to appreciate its precise role in the wage determining mechanism, and make any further judgement on its significance for the central wage control function.

At nationalisation the NUM was anxious to narrow the pay differential between pieceworkers and daywagemen and for most of the post-nationalisation period pay policy has been orientated towards that objective. There has been some doubt as to its effectiveness in this respect, however, and particularly so before the introduction of the NPLA, one justification for which was that it had been difficult previously to control pieceworkers' earnings so as to alter their relationship with those of daywagemen. Academics have also been involved in a long debate as to whether increases in pieceworkers' earnings led, or followed, those of daywagemen. In the absence of alternative published data, most students who have examined the pieceworker-daywage differential in mining have simply used EMS at the coalface as a proxy for pieceworkers' EMS movements, and surface EMS as a proxy for daywagemen's EMS. As well as presenting a slightly distorted view of the aggregate wage differential between pieceworkers and daywagemen as a whole, this 'next best' procedure ignores the contribution of earnings movements among pieceworkers and daywagemen elsewhere underground to aggregate wage changes. It also hides some illuminating differences between the behaviour of daywage earnings at the surface and elsewhere underground on the one hand, and between pieceworkers' earnings at the face and elsewhere underground on the other.

Figure 6.1 enables us to trace the implications of wage movements among the major groups of miners for internal wage differentials.

There can be little doubt that pieceworkers proved a serious problem for the central wage negotiators during the early years of nationalisation and during periods of national wage restraint in particular – during the wage freeze of 1948–50, for example, pieceworkers' EMS increased by an average of 4.5% per annum while the EMS of daywagemen increased by only 1.6% – mainly, it would seem, because of nationally agreed elements which, being dependent

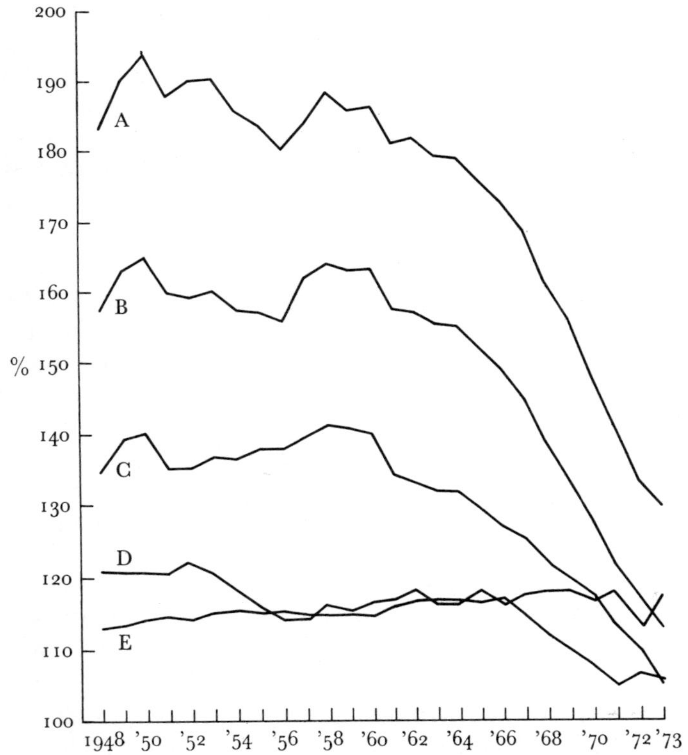

Figure 6.1 Coalmining wage differentials 1948–73

A = Face pieceworkers/surface daywagemen[a]
B = All pieceworkers/all daywagemen[a]
C = EUG pieceworkers/EUG daywagemen[a]
D = EUG daywagemen/surface daywagemen
E = Face pieceworkers/EUG pieceworkers
[a]Including underofficials and weekly paid industrial staff.
Source: NCB, Financial Returns.

on individual decisions on attendance, were beyond the reach of the central authorities to regulate absolutely. Following as it did the already growing relative wage advantage of pieceworkers during the war (see Chapter 2), these early trends seem to have had a repercussive effect on the daywage structure and the picture thereafter is one of a see-saw relationship between these two major groups of miners. With the exception of 1952, daywagemen's EMS rose faster than that of pieceworkers until 1956. Then for two years, 1957–58, when there was a nationally agreed standstill on national and local rates, pieceworkers' EMS rose faster than that of daywagemen. Thereafter, pieceworkers' EMS rose in every year, less fast than that of daywagemen. And, while by far the largest contraction in the pieceworker/daywage differential followed the national revision of the face-pieceworker structure in 1966, wage policy during the late 1960s and early 1970s merely accentuated what was an already declining relative wage advantage of pieceworkers in the previous eight years. The differential fell from 189% in 1958 to 174% in 1966; by 1973 it had been reduced to 130%.

Our data also reveal that, although face-pieceworkers' EMS increased slightly faster on average than did that of elsewhere underground pieceworkers, the difference was marginal until 1966. Thus, there would seem to have been a close association between the rates of increase in EMS among the two groups of pieceworkers before the NPLA. After 1966, the NPLA served to narrow the differential quite substantially.

The national wages policy as applied to daywagemen should have produced a progressive narrowing of the wage differential between daywagemen working underground and their lower-paid colleagues at the pit surface. Such narrowing as occurred, however, was concentrated in the period 1953–56, and most of it before the daywage revision of 1955. What is surprising is that although national wage awards served to contract the nationally agreed wage-rate structure after 1957, the actual differentials between surface and underground daywagemen increased slightly until 1971. Insofar as daywagemen were concerned, therefore, it would appear that the flat rate and ceilings policy applied to the wage structure in the three years up to 1955 was as effective in narrowing aggregate differentials as the daywage reform itself. It could also be that after 1955, the wage structure of daywagemen underground was more prone to local manipulation than was the wage structure of surface daywagemen.

We have already discussed some of the difficulties of defining the concept of wage drift in the coalmining industry – particularly in the case of pieceworkers. Even so, this analysis suggests that our residual wage items do contain an element of drift, which served to distort the wage structure from the path intended by the national bargainers. Table 6.5 gives an estimate of the contribution which various wage components made to the annual average percentage changes in EMS of the major sub-groups in various periods.

The figures in this table are calculated according to the principles outlined earlier. In this table, though, the figure for national agreements excludes the amounts payable for holiday and sickness pay and allowances in kind. It now only comprises payments for the five-day-week bonus, the proportionate impact of increases in the minimum and standard wage rates and, during the early years of nationalisation, the various flat-rate additions to pay, such as the War Addition. After 1966, the national agreement heading in the case of face pieceworkers and taskworkers includes increases in the NPLA rate. That heading also includes an estimate of the increases given under the 1971 agreement and the preceding years in the case of elsewhere underground pieceworkers and taskworkers.

If we concentrate on daywagemen first, the composition of EMS increases over the sub-periods confirms some of our earlier contentions. First, increases in the nationally agreed wage elements alone should have reduced the wage differential between surface and elsewhere underground daywagemen. Second, the higher incidence of overtime among surface daywagemen should have endorsed that reduction. But the residual wage item was much higher among elsewhere underground daywagemen (about double that for surface daywagemen after the mid 1950s) and this helped offset the influence of nationally agreed wage items (including overtime) and in some periods actually widened the differentials. However, the two periods of largest differential expansion were 1957–63 and 1966–70. These were periods in which wage ceilings were raised and larger base rate increases given to underground craftsmen. While these decisions are unlikely to explain all the discrepancies in the residual items between daywagemen underground and the pit surface, or all of the residuals in either case, they were, nevertheless, a contributing factor as were wage-structure alterations in the early 1970s. And, in the period 1957–60, the EMS of elsewhere underground daywagemen was also raised by approximately 1% per annum, due to the transference of some lower-paid pieceworkers to the new daywage structure. This also had the effect of artificially raising the average increase in EMS of pieceworkers both at the face and elsewhere underground.

These variations in the rate of wage increase between surface daywagemen and daywagemen working underground explain the different behaviour of wage differentials between face pieceworkers and

Table 6.5. *The composition of percentage wage changes among selected occupational groups 1948–73*

	Face pieceworkers				Elsewhere underground pieceworkers				Elsewhere underground daywagemen				Surface daywagemen			
	EMS	NA	OT	R	EMS	NA	OT	R	EMS	NA	OT	R	EMS	NA	OT	R
1948–51	5.4	2.2	0.5	2.8	4.9	2.4	0.5	1.9	4.8	3.6	1.0	0.2	4.6	3.9	1.1	−0.4
1951–54	6.7	0.9	0.5	5.3	6.4	1.0	0.6	4.8	6.1	4.4	0.5	1.2	7.1	5.0	1.1	1.0
1954–57	7.7	1.5	0.5	5.8	7.9	1.2	0.7	6.0	7.1	5.6	0.6	0.9	8.2	6.6	1.4	0.2
1957–60	2.2	0.8	−1.1	2.5	2.3	0.9	−1.4	2.8	2.2	1.2	−1.0	2.0	1.7	2.0	−1.3	1.0
1960–63	4.4	0.6	0.3	3.4	3.8	0.5	0.5	2.8	5.9	3.5	0.7	1.7	5.8	3.8	0.8	1.2
1963–66	3.8	0.4	0.3	3.1	3.9	0.4	0.4	3.1	5.2	3.6	0.6	1.0	5.3	4.2	0.9	0.2
1966–69	2.3	1.5	0.6	0.2	4.4	3.0	0.7	0.8	6.6	3.9	0.4	2.3	5.8	4.2	0.7	0.9
1969–73	10.8	10.5	0.5	−0.2	11.8	11.6	0.5	−0.3	15.5	12.7	0.2	2.6	15.9	14.0	1.0	1.0

Notes: The figures have been rounded to the nearest decimal point so may not sum to the total EMS figure in every case.
EMS = Earnings per manshift excluding holiday and sickness pay and allowances in kind.
NA = Nationally agreed wage items (minimum and flat rates and five-day-week bonus).
OT = Overtime premium.
R = Residual wage item.
Source: Calculated from NCB, Financial Returns.

surface daywagemen on the one hand, and between elsewhere underground pieceworkers and elsewhere underground daywagemen on the other. In the mid 1950s, for example, the narrowing of the daywage structure caused the differential between face pieceworkers and surface daywagemen to narrow, while that between elsewhere underground pieceworkers and elsewhere underground daywagemen widened.

It is important to appreciate that the pieceworkers' residual reflects the sum total of EMS increases derived from local and district bargaining, and that this was the main source from which they could get pay increases. Part of the pieceworkers' increases over the period undoubtedly exceeded the intentions of the national negotiators. But, apart from increases during the wage freeze of 1948–50, it is difficult, without a more thorough analysis of wage behaviour in the individual coalfields, to assess what proportion of the increase in the pieceworkers' residual item represented an evasion of national policy. Pieceworkers' wages were linked to productivity gains. As productivity was virtually static between 1950 and 1956, one might conclude, on that criterion, that, apart from the flat-rate addition given to pieceworkers in 1952, all the pieceworkers' wage increases over that period represented an evasion of national policy; we have already observed, for example, that there was widespread manipulation of allowance payments in the early 1950s. Nevertheless, this was given some authenticity by district officials who, in the absence of a central agreement to revise price lists, without which pieceworkers' earnings were effectively frozen, were increasingly using earnings norms in wage negotiations and raising pieceworkers' earnings by locally agreed increases in allowances. And if pieceworkers' wage rates had remained static they would eventually have been overtaken by the minimum wage rate and thereby raised by national agreement. By the mid 1950s, new agreements for power-loading operations were ratified by the national negotiators, so wage increases from this source can hardly be thought of entirely as wage drift. The widening of the piecework-daywage differential between 1956 and 1958 was partly the consequence of these agreements, as well as the increased productivity which followed the new machines.

One interesting point to emerge from the table is that before 1966, and in the mid 1950s and early 1960s in particular, the size of the residual wage item of pieceworkers – which we take as a proxy for the contribution of local and district wage bargaining to the pieceworkers' wage increase – corresponded fairly closely over the relevant periods to that of the nationally agreed wage element in the wage increase of daywage workers. Thus, over some three or four years, there was a close correspondence between the results of local and district bargaining on the one hand and national bargaining on the other. This impression is endorsed by a statistically significant positive association between annual changes in the nationally agreed minimum wage rate and changes in the residual wage item for pieceworkers ($r = 0.68$). There was also a positive association between changes in the nationally agreed minimum wage rate and changes in the residual item for daywagemen ($r = 0.56$) over the period 1948–66. But this could reflect the inclusion of some nationally agreed elements in the residual item.

One reason why the NCB and NUM preferred to express national wage increases to daywagemen in flat-rate money terms, rather than as percentage increases, was a hope that these money amounts would be used as a standard in piecework bargaining and that norms based on the flat rate would help to contract the piecework/daywage differential. Although the increase in the piecework residual was certainly less than the increase in nationally agreed wage items to surface daywagemen over the period 1962 to 1966, pieceworkers, and face pieceworkers in particular, do not seem to have succumbed wholly to this intended money illusion. Thus, the differential between them and daywagemen did not contract to the extent presumably intended by the national negotiators. With face pieceworkers' average EMS some 80% above that of surface daywagemen in 1957, for example, a national wage-rate increase to raise surface daywagemens' EMS by 3.5% per annum in the early 1960s would have implied a 2.2% increase for face pieceworkers, within which one would have had to reward increases in productivity. Even so, an average rate of EMS increase for face pieceworkers which was less than that for surface daywagemen after the late 1950s, and substantially less than a growth of productivity of approximately 5% per annum between 1957 and 1965, was surely something of an achievement for central wage regulation, particularly because, as we shall show in Chapter 8, it was occurring against a generally deteriorating wage relativity between miners and workers in other industries.

Perhaps the most ironical aspect of the analysis is that, whereas local piecework bargaining was undoubtedly a problem in the early 1950s, the wage differential contraction between pieceworkers and daywagemen which occurred thereafter was, until 1966, effected in part, by what appears to have been a wage drift, and a higher incidence of overtime working, among daywagemen than among pieceworkers. This might be seen as an attempt by daywagemen to keep their earnings rising in step with those in other sectors. We shall have more to say on this in Chapter 8.

Our table finally reveals the very great success of the national wage structure revisions of 1966 and 1971 in subsequently regulating the rate of EMS increase of miners who had hitherto been pieceworkers and taskworkers. Nationally agreed rates under these agreements were almost certainly effective in the place of work, although overtime did leave some scope for local manoeuvring. And it is interesting that the value of the overtime premium for men whose wages were previously governed by local and district price lists, at the coalface and elsewhere underground, increased quite substantially in the late 1960s while it fell in the case of all other miners over the same period. This reflects a switch from local bargaining on piece rates and allowances to overtime as a method of raising pay in the late 1960s. Thereafter the cost of overtime rose substantially with larger national awards and it was more rigorously controlled.

Why did faceworkers accept such a substantial reduction in their relative wage advantage over other grades? Following Turner,[21] we might look for evidence to suggest a decline in their political influence in the NUM. There was certainly an increased proportion of craftsmen on lodge committees by the 1960s, although otherwise the evidence in terms of political representation is slight. But this is to dismiss political influence too lightly, if one accepts that power in wage bargaining tends to derive from the product market. Coalmining is a good example of a sudden change in the product market being absorbed and reflected in the internal wage structure.

Pieceworkers constituted approximately 40% of the industry's labour force but were responsible for some 70% of the industry's wage bill in the late 1950s. This made them a particularly attractive, and acceptable, group on whom to economise. They were also the group whose power in wage negotiations was most directly related through the wage system to the fortunes of particular coalfields and pits within them. At a time of falling coal demand and increased price competition that power was severely eroded, and particularly so in the marginal coalfields, where they would receive little sympathy should their actions be seen to cause the advance closure of pits. Thus, after the late 1950s, pieceworkers became increasingly amenable to the NCB's more rigorous wage control procedures which eventually included the abolition of piecework.

It is not surprising that a contraction of an internal structure of the magnitude described here created pressures for some restoration of internal differentials. That pressure finally broke through in the national negotiations of 1975, against a background of falling productivity, the need to raise coal output and a reluctance of workers to volunteer for facework, but it was soon checked by the reimposition of incomes policy in July 1975 and for the next two years internal wage differentials shrank still further in percentage terms. It is predictable, however, that with a more buoyant product market, pressure will be exerted to widen the internal wage structure.

6.7 SUMMARY AND CONCLUSIONS

Further comment on the impact of national wages policy on wage movements and the latter's determination must await an examination of wage behaviour in the various coalfields. Nevertheless, our analysis so far permits some general observations on the role of central wage decisions in wage movements and the problems confronting the national bargainers.

Previous attempts to estimate the impact of national wage decisions on coalmining wage movements have given a misleading impression of their importance. A large part of what other writers have estimated as wage drift, by comparing the relative movement of nationally agreed and actual wage rates, was accounted for by the growth of wage items which, although provided for by national agreements, are not included in the estimates used for nationally agreed wage rates, namely the minimum wage. Such wage drift as remained using the conventional method of measurement was considerably smaller than earlier estimates, and by the early 1960s actual wage rates were, in fact, increasing less fast than our revised figure of items derived from national agreements.

It may be interesting, and useful for predictive purposes, to compare percentage changes in actual wage rates and the items of which they are composed. But such comparisons may give – as in coalmining – a misleading impression of the effectiveness of central wage decisions in regulating wage movements, unless the assumptions which underlie them correspond to the actual intentions of the national negotiators. Only by identifying the elements from which wage increases are compounded, and the relationship between them, in the light of wage policy objectives, can we understand the role of central wage decisions on wage movements and structure in the coalmining industry since nationalisation, as well as the problems confronting the central bargainers. This is a point to bear in mind when comparing wage drift between sectors.

Coalmining wage movements over much of the period considered have not been the product of what are often described as nationally agreed wage rates on the one hand, and so-called wage drift on the other: they have been the product of a number of wage elements, many of which have been payable under national agreements but whose absolute size is dependent, in part, on factors such as absenteeism, sickness, available working hours (and the willingness to work them) and possibly productivity, which national agreements cannot, and were never intended to, wholly determine. Local wage bargaining has certainly played an important, but diminishing, role in aggregate wage movements. But its major impact has been on the wages of workers for whom it was intended as a method of regulating pay.

Furthermore, as variations in particular wage components seem to both affect and reflect variations in the remainder, an explanation of changes in any one of them which ignores changes in the others gives an erroneous picture of the wage determining process in coalmining. Movements and changes in the relative importance of the pay components have often reflected different but, to a varying extent, related methods of arriving at a particular result.

In the longer term, the industry's wage structure has moved in accordance with the professed egalitarian objectives of national wages policy. Nevertheless, the ability to effect that objective, and thus its role in shaping internal wage structure, has been by no means uniform over the period considered, or, as we shall see shortly, in any coalfield.

Pieceworkers, and especially those working at the coalface, did pose a particular problem for the central wage administrators until the late 1950s. And local wage revisions in their case may have weakened central authority over daywages. Such local manoeuvring as there was, however, was confined mainly to daywagemen working in close proximity to pieceworkers underground. The differential between pieceworkers and daywagemen contracted for brief periods but the see-saw relationship implied some resistance to contraction over the medium term. However, the differential increased mainly in those periods when basic wage rates increased least or were frozen. Thus, although a slower growth of EMS was achieved in the short term, it was at the cost of some internal differential distortion which was then corrected in subsequent national negotiations.

The short-lived widening of the internal wage structure between 1956 and 1958 coincided with a labour regrading exercise and a rapid increase in the use of power-loading machines. Thereafter the differential between pieceworkers and daywagemen narrowed as piecerate revisions failed to keep up with increases in nationally agreed wage rates, the wage ceilings policy was relaxed, and daywagemen increased the number of hours they worked in overtime. This narrowing process accelerated rapidly after the national revision of the piecework structure in 1966 and 1971.

Although it was consistent with central wages policy, the narrowing of internal wage differentials after the 1950s can be seen as an adjustment to a sudden collapse in the product market. One would predict, therefore, that the more recent, but equally dramatic, strengthening of coal's market position as a result of increased oil prices will act as a powerful incentive to widen the internal wage structure in the medium term.

7. THE COALFIELD WAGE STRUCTURE

In the previous chapter we outlined the mechanics of wage determination and the impact of national wage decisions on aggregate wage movements and wage differentials. A full appreciation of the problems confronting the central wage negotiators, as well as the impact of their decisions on wage levels and movements, and the latter's eventual determinants, inevitably requires a more thorough examination of wage behaviour in the various coalfields. In this chapter, therefore, we extend our appraisal of wages policy with an analysis of inter- and intra-coalfield wage relationships.

7.1 Wages policy and regional diversity

We outlined in Chapter 2 some of the problems confronting the NCB and NUM in wages policy formulation at nationalisation. The NUM wanted a more uniform structure between men working in different coalfields, but realised the threat this might pose to its own organisation if an attempt was made to achieve it at the cost of the higher-paying areas. The nationalised industry was allowed to cross-subsidise its operations, but it is doubtful also whether the NUM could wholly ignore the possible implications for future employment of raising wage costs fastest in the already high-cost areas, which were also the least profitable on average, when the industry was instructed to base future investment decisions on the relative profitability of the coalfields. The NCB was also unclear as to how far it should, or could, go in the direction of wage uniformity, given the need to avoid dislocating production and its reliance on wage incentives to increase output.

To what extent, then, did these same considerations subsequently make for differences in the ability to effect national agreements and decisions in the various coalfields? Would not the higher-paid areas which were also the most productive and profitable be more likely to find ways of circumventing national decisions if these were found to be incompatible with local circumstances and traditions than the low-paid, but high-cost, areas, which might be concerned that the more rapid rate of wage advance which wage equalisation implied in their case threatened long-run employment security? If economic considerations were important in influencing the course of national wages policy, would they not also be important in influencing local reactions to national wage decisions when the latter seemed at odds with local priorities? One might have expected the falling demand for coal to put extra pressure on the high-cost coalfields and exacerbate the conflict between egalitarian ideals and economic priorities. And to what extent did such problems, if they existed, affect all workers alike? We have seen already that the differential between daywagemen and pieceworkers veered in different directions during various phases of the national wages policy before 1966. Was this pattern uniform for all coalfields or does it reflect significant differences between the regions? And to what extent has overtime been used to circumvent national decisions? Here, then, are some of the questions to which we address ourselves in this chapter.

7.2 Wage movements and structure 1946–66

From an institutional viewpoint, wages policy since 1946 has passed through three stages. The first ended with the 1955 Revision of the Daywage Structure. The second lasted to 1966 and the conclusion of the NPLA. The third ended in 1970 with the Third Daywage Structure.

Because of changes in geographical boundaries, it is impossible to use data which refer to any one administrative area for the whole period. The most convenient and realistic way around this problem is to use data which before 1966 refer to the wage districts into which the industry had traditionally been divided for collective bargaining purposes[1] and thereafter to the new administrative areas, and consider wage relationships after 1966 separately at the end of the chapter. The EMS statistic used in this chapter excludes holiday and sickness payments, and the imputed value of allowances in kind, and is thus the same as that used for our earlier analysis of wage differentials.

In each wages policy phase, national wage agreements were biased in favour of the lowest-paid workers and coalfields. If effective, this policy would imply percentage rates of wage advance inversely related to base period wage levels and thus, *ceteris paribus*, a progressive narrowing of regional wage differentials in percentage terms. Table 7.1 gives the coefficient of correlation between percentage increases in EMS in the 18 wage districts, over the period 1946 to 1965, and selected sub-periods chosen to cover wages policy phases (given on the vertical axis) and actual EMS levels in base periods (given on the horizontal axis). All the correlation coefficients are negative and statistically significant at either the 1% or 5% level,

Table 7.1. *Coefficients of correlation between EMS in various base years and percentage changes in EMS over selected periods 1946–65*

	Base years		
	1946	1954	1958
Periods of wage change			
1946–54	−0.61**		
1954–62		−0.53*	
1958–65			−0.49*
(Whole period)			
1946–65	−0.79**		

**Significant at 1% (*t* test).
*Significant at 5% (*t* test).

confirming that EMS tended to rise least fast in those coalfields with the highest EMS in 1946 and subsequent base years. The appropriate coefficient for the period 1946–65, for example, was −0.79.

But, while the correlation coefficients indicate a general contraction in the wage structure, that trend hides some significant exceptions. We can identify some of the more obvious anomalies with the help of Figure 7.1 where, on the horizontal axis, we have the average EMS paid in 1946 and, on the vertical axis, the percentage increase in EMS between 1946 and 1965. The coalfields are thus arranged moving from left to right across the page in ascending order of their position in the 'wage league' in 1946. The percentage increase in EMS between 1946 and 1965 is divided between that obtained in the period 1946 to 1954 and that awarded between 1954 and 1965.

EMS in Shropshire, Lancashire and Cheshire, Northumberland and the North East increased faster between 1946 and 1965 than did EMS in Durham, which was a relatively lower-paying coalfield in 1946. Again, EMS increased less rapidly in Cumberland than in North Staffs, although they were identical in 1946; and EMS in both those coalfields rose by a smaller percentage than would appear warranted by a comparison with wage movements in the South East, North Derbyshire and Warwickshire coalfields, where again the increase in EMS was not much below that in much lower-paying Durham. At the other extreme, EMS in Cannock Chase, a relatively low-paying coalfield in 1946, increased far more than in coalfields paying similar wages in 1946. But the seemingly 'anomalous' rates of wage increase in Cannock Chase, Shropshire, Lancashire and Cheshire over the years 1946–65, as well as that in Cumberland *vis-à-vis* North Staffs,

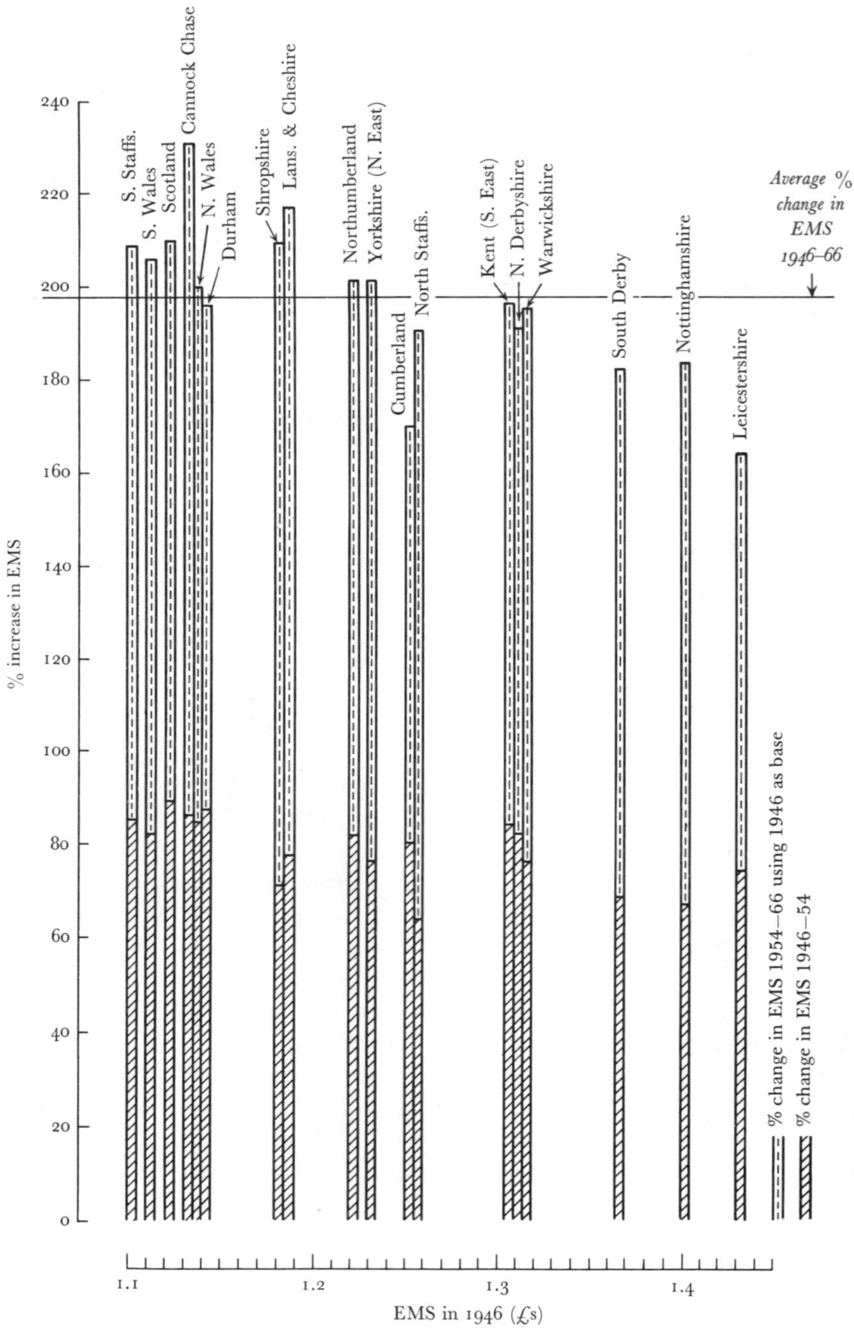

Figure 7.1 Percentage change in Coalfield EMS 1946–66
Source: NCB, Financial Returns.

would appear to be primarily a product of wage movements after 1954. What appears as a rather generous rate of wage advance in the South East, North Derbyshire and Warwickshire coalfields over the period, however, was the product of wage movements both before and after 1954. Wage movements in the three highest-paid coalfields in 1946 – South Derbyshire, Nottinghamshire and Leicestershire – are of some interest in that Leicestershire appears to have had an abnormally high rate of wage advance between 1946 and 1954, but after 1954 compared unfavourably with the neighbouring coalfields of Nottinghamshire and South and North Derbyshire. The fact that the coefficients of correlation given in Table 7.1 are smaller, the more recent the time period considered, suggests rather more exceptions to miners in the lower-paying coalfields getting larger percentage increases than miners in the higher-paying coalfields with the passage of time. The fact that the coefficient for the period 1946–65 is higher than for any one sub-period may indicate see-sawing in the relative rate of increases in the short run.

We shall return to these somewhat anomalous wage movements later. First, let us examine the implications of the above trends for regional wage differentials. Table 7.2 gives, for selected years, the percentage by which average EMS in each wage district was above (or below) the unweighted average of the coalfields as a whole, the percentage and absolute range of EMS between the highest- and lowest-paying coalfields, as well as the coefficient of variation. In 1946, for example, average EMS was highest in Leicestershire, which had a wage advantage of 27.4% over South Staffs, the lowest-paying coalfield. Taking all coalfields together, the coefficient of variation was 8.4%.

By 1966, the coefficient of variation had fallen to 5%: a fall which was associated with a sharp reduction in the percentage range of EMS between the highest- and lowest-paying coalfield. But, whereas there was a general trend towards greater wage equality, this was by no means uniform over the periods considered. Thus, between 1954 and 1958, Nottinghamshire, the highest-paying coalfield, actually *increased* its wage advantage, while the relative position of South Wales, one of the lowest-paying coalfields, actually deteriorated. These contrary trends may help to explain why the percentage range of EMS *increased* between 1954 and 1958, while the coefficient of variation decreased from 6.7 to 6.0%.

Data for 1938[2] are given purely for comparative interest. They reveal a larger wage structure contraction between 1938 and 1946, when wages were adjusted by flat rate cost-of-living increases, than in any subsequent eight-year period before 1966. Between 1946 and

Table 7.2. *Coalfield wage district differentials 1938–66*

	District wage indices; unweighted average = 100					Rank order				
Wage district	1938	1946	1954	1958	1966	1938	1946	1954	1958	1966
Leicester	107.1	117.1	109.8	105.1	103.5	6	1	2	4	6
Notts.	126.3	114.6	111.6	111.9	109.0	1	2	1	1	1
S. Derby	94.7	111.7	105.7	103.6	105.7	11	3	5	7	4
Warwick	119.2	107.3	105.7	104.3	106.0	2	4	6	5	3
Kent	107.3	106.5	109.8	110.1	106.3	4	5	3	2	2
N. Derby	107.2	106.5	108.4	105.9	103.5	5	6	4	3	5
N. Staffs	101.8	101.6	91.6	94.2	99.2	7	7	18	13	11
Cumberland	94.6	101.5	102.5	94.1	92.4	12	8	7	14	16
Yorkshire	110.5	100.0	93.9	102.2	100.8	3	9	9	8	9
Northumberland	85.9	99.2	100.7	104.0	100.0	17	10	8	6	10
Lancs. and Cheshire	94.8	95.9	95.2	98.9	102.2	10	11	13	9	7
Shropshire	86.0	95.8	92.0	93.9	98.5	16	12	16	15	12
Durham	84.2	94.3	96.6	98.6	92.4	18	13	10	10	17
Cannock Chase	93.0	91.9	95.7	96.4	101.6	13	14	12	12	8
N. Wales	91.2	91.8	94.8	93.1	92.1	14	15	14	17	18
Scotland	94.9	91.6	96.1	98.2	97.5	9	16	11	11	13
S. Wales	96.5	90.2	92.0	91.7	92.6	8	17	17	18	15
S. Staffs	89.5	89.4	92.9	93.5	95.4	15	18	15	16	14
% range	42.1	27.7	20.0	20.2	16.9	Spearman's rank correlation				
Range in pounds	0.24	0.21	0.44	0.56	0.62		1938/46	0.61		
Coefficient of variation %	11.4	8.4	6.7	6.0	5.2		1946/54	0.78		
							1954/58	0.88		
							1958/66	0.86		
							1946/66	0.75		

Source: NCB Financial Returns

1966, however, the largest contraction in the regional wage structure occurred in the eight years before 1954. Although the extent to which a wage structure can contract must depend partly on the degree of disparity from which one starts (as well as on the general pace of wage advance, which was certainly smaller in percentage terms during the period 1958–66 than it had been in 1938–46 and 1946–54), this suggests that the *ad hoc* policy of flat-rate wage advances adopted up to 1954 was as effective, if not more so, in reducing regional wage disparities as the more systematic policy incorporated in, and possibly facilitated by, the revision of the daywage structure in 1955.

The rank correlation coefficients indicate that, while there was a significant contraction in the wage structure, this was not associated with a general upheaval in the coalfield wages league, and that in rank order terms the wage structure became relatively more stable with the passage of time. This is consistent with our earlier finding that the wage structure showed more resistance to contraction after the 1950s. Some of the seemingly anomalous wage trends highlighted earlier, nevertheless, produced some changes in the league table.

The Midland coalfields and Kent maintained their superior wage position throughout the period. But whereas, on average, Leicestershire miners were the highest paid in 1946, Nottinghamshire miners moved to first place by the 1950s and retained their lead through to 1966, while the Leicestershire miners slumped to sixth position in the wage league. Again, South Derby fell from third to fourth position between 1946 and 1966, thus changing positions with Warwickshire. Of those coalfields occupying between eighth and twelfth positions in 1946, Cumberland's ranking fell from eighth to sixteenth by 1966, while Lancashire and Cheshire moved up from eleventh to seventh position. The remainder of the coalfields in this middle ranking range show very little deviation from their wage position in 1946. Of the remaining coalfields, Cannock Chase improved from fourteenth to eighth position, while South Staffordshire rose from eighteenth to fourteenth. But perhaps Durham miners experienced the most startling change of fortune. By the mid 1950s, Durham had moved from the bottom to tenth position in the wage ladder. By 1966, however, Durham was back near the bottom of the wages league. And it is interesting to note that a similar, if less dramatic, reversal of fortunes occurred in Northumberland, Durham's neighbouring coalfield, over the same period.

So far then, our analysis confirms our earlier finding that, whereas wages moved in a direction which was generally consistent with the egalitarian objectives of national wages policy over the period 1946–65, other influences were at work to fashion the regional wage structure. At this point, it will be helpful to examine wage structure changes among the major occupational groups.

7.3 The inter-coalfield occupational wage structure

On *a priori* reasoning we should expect most of the apparent flexibility in the wage structure to reveal itself in the wages of pieceworkers. Again, we shall distinguish between pieceworkers performing tasks at the coalface and elsewhere underground respectively, and daywagemen in jobs located elsewhere underground and at the pit surface. Table 7.3 presents data similar to those in Table 7.2, although we

Table 7.3. *Wage district differentials of workers in various payment categories (district wage indices: unweighted; average = 100)*

	Face pieceworkers			E. underg'd pieceworkers			E. underg'd daywage			Surface daywage		
	1950	1958	1965	1950	1958	1965	1950	1958	1965	1950	1958	1965
Leicestershire	112.3	106.1	104.5	127.4	106.3	102.9	107.7	100.0	103.4	114.6	103.1	101.9
South East	112.3	117.5	111.1	109.5	117.9	107.1	104.1	102.9	104.8	101.7	103.7	101.5
South Derby	111.7	104.2	107.6	115.9	112.8	108.1	102.1	98.6	102.1	107.8	101.1	102.2
Warwick	110.9	109.8	106.7	100.5	100.3	106.1	104.2	102.5	103.8	103.4	98.6	101.4
Nottingham	109.1	115.1	113.6	109.6	109.3	109.8	109.1	105.8	104.1	119.0	107.1	105.2
North Derby	105.7	108.2	107.8	104.5	104.2	102.2	106.3	103.0	100.4	102.6	101.0	101.8
Northumberland	103.9	104.8	98.4	99.5	96.7	101.2	95.8	99.0	100.6	94.7	98.9	98.9
Cumberland	102.2	92.1	94.9	104.5	98.4	92.2	101.4	98.2	96.3	94.1	95.8	95.5
North East	99.6	102.9	100.6	99.6	99.7	101.6	97.9	99.1	98.3	99.1	100.1	99.5
Shropshire	96.1	91.8	95.5	95.5	101.2	100.7	101.5	94.7	97.9	94.8	95.9	99.7
Lancs. & Cheshire	95.7	97.1	106.8	92.0	96.1	101.7	93.0	98.1	100.3	97.4	103.6	101.7
Durham	94.3	93.4	84.6	93.5	92.5	75.2	91.6	98.5	95.5	94.0	100.0	96.6
North Staffs.	93.9	96.6	98.2	93.8	95.9	101.5	97.2	97.1	99.7	98.4	99.5	100.7
South Staffs.	93.5	93.1	95.7	75.6	99.4	91.7	102.8	106.3	100.7	97.4	99.0	101.1
Cannock Chase	92.7	94.4	101.2	100.0	90.4	97.1	96.5	98.7	100.8	95.7	98.0	97.4
North Wales	92.6	88.9	92.2	96.0	92.6	91.2	90.2	96.6	96.9	98.3	98.5	99.3
Scotland	91.3	95.8	92.6	100.4	98.5	100.4	100.0	101.9	99.8	93.9	95.8	95.6
South Wales	85.2	86.5	98.5	86.1	88.4	93.4	94.4	102.4	97.6	100.0	102.0	100.0
% range	27.1	31.0	29.0	51.8	29.5	34.6	18.9	11.6	9.3	25.1	11.3	9.7
Range in pounds	0.62	1.17	1.21	1.54	0.92	0.78	0.27	0.24	0.25	0.28	0.22	0.26
Mean (unweighted) (£s)	2.30	3.78	4.86	2.01	3.35	4.07	1.43	2.08	2.91	1.16	1.96	2.67
Coefficient of variation %	8.3	8.8	7.7	11.0	7.5	5.9	5.4	3.0	2.7	6.8	2.9	2.5

Source: NCB, Financial Returns of the Wage Districts

have been unable to provide data for years before 1950. The pattern of wage behaviour produced by including data for years other than those selected varied only slightly from that portrayed by the years selected in the case of face pieceworkers.

The largest regional variation in EMS (as measured by the coefficient of variation) in 1950 was among pieceworkers working elsewhere underground (11.0%), followed by face pieceworkers (8.3%). The smallest variation was among surface daywagemen (6.8%). The percentage difference between EMS in the highest- and lowest-paying coalfields, however, ranged between 52% in the case of elsewhere underground pieceworkers, 27% for face pieceworkers and 25% for surface daywagemen. The largest contraction in wage differentials on both dispersion measures was among surface daywagemen, followed by pieceworkers and daywagemen elsewhere underground and, lagging a long way behind all these groups, face pieceworkers. However, the largest percentage difference in EMS, between the highest- and lowest-paid coalfields, was still among elsewhere underground pieceworkers. Our figures show that, whereas the larger part of the wage structure contraction among elsewhere underground pieceworkers and daywagemen generally occurred between 1950 and 1958, the face pieceworkers' wage structure actually *widened* during that period. A more detailed analysis of the period 1950–58, however, revealed that the expansion of the faceworkers' wage structure was mainly the result of wage trends between 1956 and 1958: the coefficient of variation fell from 8.3% to 7.7% between 1950 and 1955 and then increased to 8.8% in 1958.

These trends are quite consistent with the negative correlation coefficients between percentage increases and base period level of EMS over the period 1950–66 shown in Table 7.4, being highest in the case of surface daywagemen and weakest among face pieceworkers. This pattern was repeated in every sub-period except 1958–66, when the inverse association was weakest among surface daywagemen. The statistical significance of the correlation coefficient for face pieceworkers, however, is below 5% in every period except 1958–66, and was particularly weak in 1954–58. The coefficients for daywagemen are higher for surface than underground workers before 1958. The coefficients are much higher for both groups in the period 1954–58 than in 1950–54.

These results indicate that the contraction of regional wage differentials during the first decade of nationalisation was due almost entirely to wage trends among miners other than face pieceworkers, on whose wage structure the egalitarian aspects of policy apparently had little systematic impact. Our results would tend to confirm that

Table 7.4. *Correlation coefficients between EMS in various base years and percentage changes in EMS over selected periods according to wage payment system and place of work*

Periods of	Face pieceworkers			Elsewhere underground pieceworkers			Elsewhere underground daywagemen			Surface daywagemen		
wage change	1950	1954	1958	1950	1954	1958	1950	1954	1958	1950	1954	1958
1950–54	–0.36			–0.38			–0.44			–0.70**		
1954–58		–0.05			–0.66**			–0.85**			–0.88**	
1950–58	–0.16			–0.71**			–0.83**			–0.91**		
1958–66			–0.45			–0.60**			–0.58**			–0.52*
1950–66	–0.42			–0.87**			–0.87**			–0.95**		

** Significant at 1% (t test)

* Significant at 5% (t test)

Source: NCB, Financial Returns of the Wage Districts

the behaviour of face pieceworkers' wages acted as a destabilising influence on the wages of other workers underground before the reform of the daywage structure. It may seem surprising that, whereas national wage decisions seem to have had little influence on the wage structure of face pieceworkers, they do seem to have influenced that of pieceworkers working elsewhere underground in the period 1954–58. The reason for this, though, is partly statistical. Some men doing underground jobs which were subsequently brought into the new daywage structure were paid on a piece rate or, more usually, a high fixed task rate. Many of these anomalies were tidied up in the 1955 daywage revision and part of the contraction in the elsewhere underground pieceworkers' wage structure between 1950 and 1958 was probably due to piecework tasks being brought into the daywage structure. In addition, though, it could be that the NCB was having greater success, after the 1955 daywage reform, in imposing the pattern of daywage increases on elsewhere underground pieceworkers, many of whom were on enhanced day rates, which tended to be adjusted with changes in the daywage rates of miners working alongside the enhanced-rated men. If this was the case, though, the success clearly did not extend to face pieceworkers until the late 1950s and early 1960s.

Table 7.3 enables us to elaborate on some of the seemingly anomalous coalfield wage movements referred to earlier. For example, Durham's slump from the middle to near bottom of the wages league was mainly due to the deteriorating wage position of pieceworkers in that coalfield. The improvement in Lancashire's wage position, on the other hand, reflected an above average wage increase for each of its wage groups; although there, too, pieceworkers played a prominent role. Thus, whereas face pieceworkers in Lancashire and Durham were on virtually identical EMS in 1950, the EMS of Lancashire face pieceworkers was approximately 22% above that of their Durham colleagues in 1965. Face pieceworkers' earnings seem to have been the main reason for Nottinghamshire's more favourable average wage increase compared with Leicestershire; although the failure of other groups in Leicestershire to keep pace with the industry average was also a contributing factor. With the exception of wage trends in the coalfields cited above, however, the correlation coefficients given in Table 7.5 show the rank ordering of the differentials in Table 7.3 to have been remarkably stable.

As the degree of stability in the rank order is likely to be inversely related to the degree of wage dispersion in the base year, we should not, in the light of the relatively small overall wage dispersion among daywagemen compared with pieceworkers, attach too much signific-

Table 7.5. *Spearman's rank correlation coefficients of district EMS of workers by payment category at the coalface, elsewhere underground and the pit surface 1950–65*[a]

Correlated years	Coalface pieceworkers	Elsewhere underground		Surface daywagemen
		Pieceworkers	Daywagemen	
1950/58	0.81	0.72	0.62	0.67
1958/65	0.87	0.78	0.62	0.78
1950/65	0.75	0.68	0.72	0.86

[a]Calculated from Table 7.3. All the coefficients are statistically significant at the 1% level.

ance to the minor differences in the size of the rank correlation coefficients between various groups.

One further point to emerge from Table 7.3 is that coalfields paying above-average wages to one work group tended to pay above-average wages to the other groups. For example, even in 1965 there was a rank correlation coefficient of 0.78 between the EMS of face pieceworkers and of surface daywagemen. This would not be surprising had the NCB and NUM not reformed the daywage structure in 1955. And in Nottingham (the highest-paying day-rate coalfield), for example, the EMS of surface daywagemen was still 9.7% above that in Cumberland. We shall return to this point shortly.

7.4 Intra-wage-district occupational differentials

The different behaviour of the wage structure of the various payment categories inevitably reverberated on intra-wage-district skill differentials. This aspect of the wage structure is illustrated in Table 7.6, where the EMS of pieceworkers is expressed as a percentage of that of surface daywagemen in each wage district. Two things emerge from this table. First, there was, in each year, a wide disparity in the differential: in 1950, for example, there was a 50 percentage points difference between the differential in the Kent (South East) and South Wales coalfields. Second, although there was a general tendency for the differential to contract between all but one of the subperiods (1954–58), the pace of contraction varied considerably. Thus, between 1950 and 1965, there was very little contraction (under 10% points) in four districts (South Wales, Cannock Chase, Leicestershire, Lancashire and Cheshire); a fairly substantial contraction (10–20% points) in eight districts (South East, South Staffs, North Staffs, South Derby, North Derby, North Wales, North East, and Scotland); a large contraction (over 20% points) in five districts (Warwicks,

Table 7.6. *Face pieceworkers – surface daywage differentials 1950–65 (per cent)*

Wage Districts	1950 (*a*)	1954 (*a*)	1954 (*b*)	1958 (*b*)	1962 (*b*)	1965 (*b*)
South East	219	213	215	219	202	199
South Wales	169	168	169	164	163	160
Warwicks.	213	211	213	215	194	191
Shropshire	201	183	184	185	181	174
South Staffs.	190	181	182	181	170	172
Cannock Chase	192	189	190	186	191	189
North Staffs.	189	176	178	187	184	177
Leicestershire	194	189	190	199	192	187
South Derby	206	193	193	199	194	192
Nottinghamshire	182	178	179	207	206	196
North Derby	204	199	201	207	201	193
North Wales	187	178	179	174	166	169
Lancs. & Cheshire	195	181	182	181	191	191
North East	199	191	192	198	190	184
Durham	199	180	183	180	171	159
Cumberland	216	211	212	185	177	181
Northumberland	217	203	204	204	188	181
Scotland	193	195	196	193	182	176
Average	198	190	191	192	186	182

(*a*) includes underofficials
(*b*) excludes underofficials
Source: NCB, Financial Returns

Shropshire, Durham, Cumberland, Northumberland). In one district, Nottinghamshire, the differential increased, although this was the result of wage trends in the 1950s.

As the magnitude of the differential contraction was positively related to the size of the initial (1950) differential ($r = 0.62$), national wages policy appears to have had a generally egalitarian influence on intra-, as well as on inter-wage-district differentials. Durham and Nottinghamshire provide two of the most interesting anomalies in this respect. Despite paying generally lower wages than Nottinghamshire, the face pieceworker/surface daywagemen differential in 1950 was larger in Durham than in Nottinghamshire. The national daywage policy raised day wages faster in Durham than in Nottinghamshire. But this was offset by a much slower rate of growth in face pieceworkers' EMS in Durham than elsewhere. Thus, while in Nottinghamshire the EMS of surface daywagemen was held back to a below-average increase (although still, we suggest, one which rose above that dictated by the national policy), the EMS of face pieceworkers rose by an above-average percentage. This, at least, suggests that the cost of the egalitarian policy was not borne wholly by the highest-paid miners in the highest-paid wage districts.

7.5 The determinants of wage structure

Taken as a whole, our evidence supports our earlier contention that, whereas in the longer term the wage structure moved in a manner which accords with the egalitarian objectives embodied in national wage agreements, the latter were not the only factors of importance in influencing wage behaviour in the coalfields, and thus the aggregate trends outlined in the previous chapter. In view of this, we clearly need to examine the wage structure in more detail. We have already identified several sources of flexibility. Perhaps the most important were piecework price lists. As pieceworkers' wages might appear to have posed a special problem for the national negotiators, we shall start by examining factors which seem to have influenced them.

Before we do so, though, we must issue a note of warning: because of the rapidly changing circumstances affecting the coalmining industry over the twenty years considered here, and their varied impact on particular coalfields, we would not necessarily expect coalfields to react uniformly to one of them. This necessarily makes an analysis of wage determination in mining rather complex. It is the relationship between the particular and the general, however, which makes it a fascinating subject.

7.5.1 *Piecework bargaining*

By attempting to relate wages to productivity, a piecerate system provides perfectly legitimate means whereby the wages of some workers or groups can rise faster than those of others. That they moved in this manner during the early years of nationalisation, however, there is no evidence. At nationalisation, piece rates were still frozen under the terms of the 1944 Revision of the Wage Structure Agreement and, with the exception of the 1951 flat-rate award, they received no other national wage increase.

Under the pressure of production, pieceworkers found ample room for manoeuvre during the early years of nationalisation, and in every coalfield wages rose faster than intended by the national negotiators. Between 1950 and 1955, however, the pattern of face pieceworkers' increases, influenced in part by the relative trend of daywages as well as the 1951 flat-rate addition to wages, was sufficiently biased in favour of the lower-paying coalfields to produce a narrowing of the face pieceworkers' wage structure, as well as the differential between face pieceworkers and surface daywagemen. But that pattern also accords fairly closely with the amount of pressure applied to pit managements through work stoppages and restrictions. And if we examine Table 7.7, we can also see that, at a Divisional level at least, the pattern of militancy would also explain why face pieceworkers in Kent received a relatively higher increase in EMS than would seem warranted by their initial wage rank and why Durham miners achieved less. In short, the piecerate system appears to have responded to the amount of pressure face pieceworkers were prepared to apply in the face of the differential erosion by other groups implied by national wage revisions at the time.

By the mid 1950s, new factors were serving to influence, and this time widen, the face piecework structure as well as the differentials over datal workers, who, as a result of the 1955 Agreement, had even less room for manoeuvre than before. Of these factors, the most important was the introduction of power loading to replace conventional coal-cutting equipment. Power loading affected face pieceworkers' earnings and helped increase disparities between them in three inter-related ways: first, because it was a new technique, it required new wage agreements; second, it greatly increased productivity; third, its impact was uneven between coalfields.

To encourage acceptance of the new machinery, it was generally agreed that power-loading operators should have an earnings norm slightly above that prevailing on conventional machine-cut coal faces. General acceptance of this principle would ensure that the percentage

Table 7.7. *Face pieceworkers' wage increases and pit militancy 1950–54*

	Percentage increase in face pieceworkers' earnings per manshift		Work stoppages and restrictions over wages and price lists per pit	
	%	Rank	No.	Rank
Scotland	35.24	1	2.23	1
S. Wales	35.20	2	1.73	3
Kent	31.00	3	1.45	4
Lancs. & Cheshire	30.00	4	1.17	5
Yorkshire	29.31	5	2.05	2
West Midlands	28.50	6	0.76	6
East Midlands	28.06	7	0.41	7
Northumberland & Durham	26.80	8	0.18	8

Source: NCB, Financial Returns and Dispute Returns

increases in face pieceworkers' EMS would tend to be positively related to the pace at which power loading was introduced. The spread of power loading thus helps to explain the widening of the face-pieceworker/surface-daywage differential on average in the late 1950s. And, turning to Table 7.8, the uneven pace at which it was introduced would help to explain the widening of the face pieceworkers' wage structure in the late 1950s, and in particular the above average increase in EMS in Nottinghamshire and Lancashire of which it was mainly a product.

But power-loading machines also had a substantial impact on coal face productivity, which had been virtually static for the first decade of nationalisation. To what extent, therefore, do differences in the rate of increase in face pieceworkers' earnings after 1956 also reflect *differences* in productivity movements caused by the uneven spread of power loading in the coalfields? This question can be answered, in part, with the help of Figure 7.2, which illustrates that the relatively high increases in EMS in Nottinghamshire, Lancashire, Cannock Chase, and North Derby were all associated with above-average rates of productivity growth. Again, but at the other extreme, the relatively small increase in face pieceworkers' EMS in Durham was associated with relatively small productivity gains. And, although the correlation coefficient between the two variables for all eighteen wage districts over the period is only +0.28, this rises to +0.48 if the South Staffs and Shropshire wage districts, which consisted of only a few pits, all fighting for survival, and exceptional in not increasing productivity by more than 30%, are excluded from the calculation, (this latter relationship is depicted by the thinner regression line).

But, despite a very general association between increases in face

Table 7.8. *Percentage of coal obtained by power loading*

Division	1950	1955	1960	1965
Scotland	1.0	6.5	31.9	72.4
Northern	4.5	8.1	28.6	66.6
Durham	2.8	5.5	30.8	
Yorkshire	2.8	9.8	31.7	85.6
North Western	0.8	18.0	39.6	93.4
East Midlands	10.8	16.4	60.5	91.5
West Midlands	2.7	8.1	27.8	86.6
South Wales	0.6	4.0	23.4	57.8
Kent	0.0	1.2	10.9	84.7

Source: *Digest of Energy Statistics*, 1963 and 1967

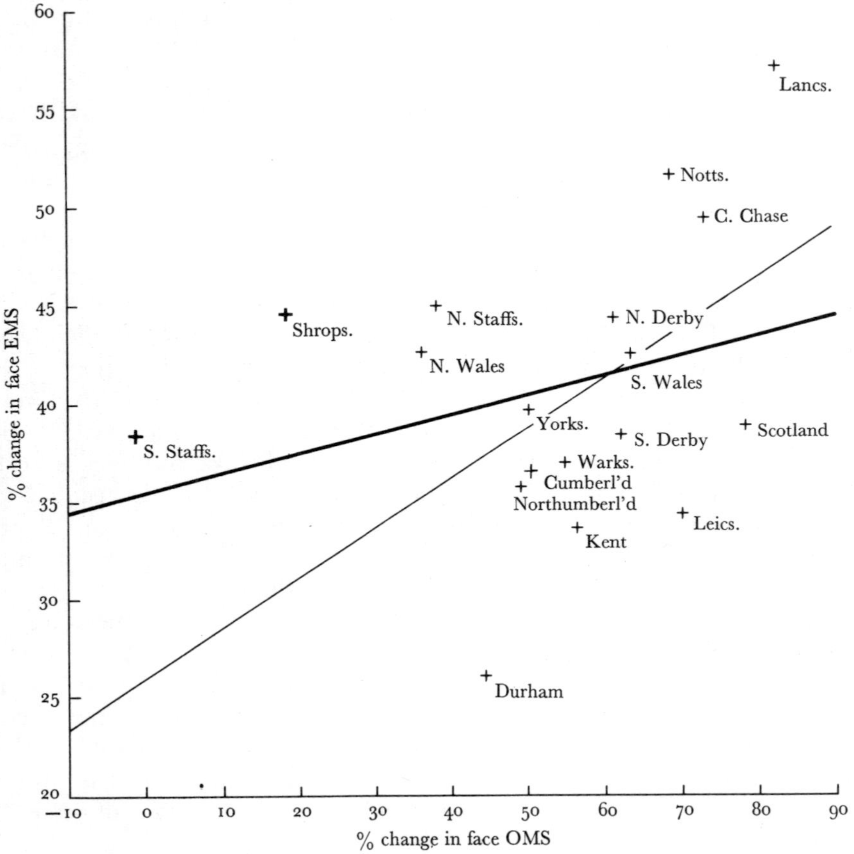

Figure 7.2 Percentage change in face pieceworkers' EMS and face productivity 1956–65
Source: NCB, Financial Returns.

pieceworkers' wages and productivity, the association gave no evidence of marked regularity. Thus in Scotland and Leicestershire the rates of productivity advance were not dissimilar to those in Lancashire and Nottinghamshire, although in the former coalfields the growth of face pieceworkers' earnings was much smaller. At the other extreme, there were several coalfields where productivity growth was not much better than that in Durham but where face pieceworkers' earnings rose much faster.

Some of this seemingly anomalous behaviour is probably explained by the nature of the power-loading agreements and the circumstances of particular coalfields. Let us deal first with the agreements.

We have remarked already on the inadequacies of central wage control before the restructuring of the industrial relations function following the Fleck Report. It was symptomatic of that weakness that power loading was allowed to progress without a systematic central blueprint for wages on such machines. In this situation it would have been surprising indeed if the new wage arrangements did not differ substantially between the coalfields and, should the two conflict, reflect local (rather than national) considerations.

In Lancashire, where there had been a long history of local autonomy in wage matters and the local lodges were particularly strong, power-loading operations were covered by pit-piecework agreements. This was also the case in Kent, until 1965, when officials in that coalfield opted for a district agreement. In all the other coalfields, district agreements were the fashion, in an attempt to introduce greater uniformity into wage provisions at the coalface and subsequently enhance the power of the district officials. But even so the agreements differed between coalfields. Yorkshire officials decided to offer a choice of wage agreements to their power-loading teams, and by 1960 half of them had opted for the piecework-type agreement and half for the daywage. In Durham and Scotland, where seam conditions were relatively poor and the lost earnings through mechanical breakdowns particularly high, task rate agreements, under which men were paid a fixed daywage, plus a fixed bonus in the event of their completing a predetermined task, were negotiated. The piecework agreements adopted in the other coalfields provided for a fairly high fall-back rate, but the variable wage element in earnings was potentially higher in those agreements than in the case of the task rate agreements of Scotland and Durham.

The piecerate agreements would naturally relate wage increases more closely to productivity changes than would the daywage-style agreements, although the actual rates of increase might, in either case, also depend on the pressures brought to bear in negotiation. Equally important, piecerate agreements usually tended to offer greater scope for manipulation. In this context, it is especially worth noting that increases in wages lagged a lot further behind productivity increases in Scotland and Durham, where daywage-type agreements operated for power-loading faces, than they did in coalfields with piecerate agreements. At the other extreme, Lancashire face pieceworkers, who had the higher rate of wage advance in this period, had local piecerate agreements as well as the highest productivity.

It could be, of course, that the wage/productivity relationship really worked the other way round, and that differences in productivity were caused, in part, by the style of agreement adopted, i.e.

productivity rose fastest where the wage incentive was most closely related to output. Such comparisons as that between productivity changes in Lancashire and Scotland do not really support this argument – nor does the interesting comparison of wage and productivity movements in the four wage districts which constituted the East Midlands coalfield: Nottinghamshire, Leicestershire, North Derby and South Derby. Whereas there was a difference of only eight percentage points in the rate of productivity increase in these districts over the decade, there was a difference of nearly twenty percentage points in the growth of face pieceworkers' EMS. And as these latter wage districts were on similar piecework agreements, the differences in wage increases seem explicable only in terms of local factors affecting the wage control function. As Nottinghamshire, in particular, has proved a serious anomaly in our analysis to date, it is worth examining these problems a little further here, as a more general illustration of the factors influencing wage movements in particular coalfields. This we can do with the help of Figure 7.3.

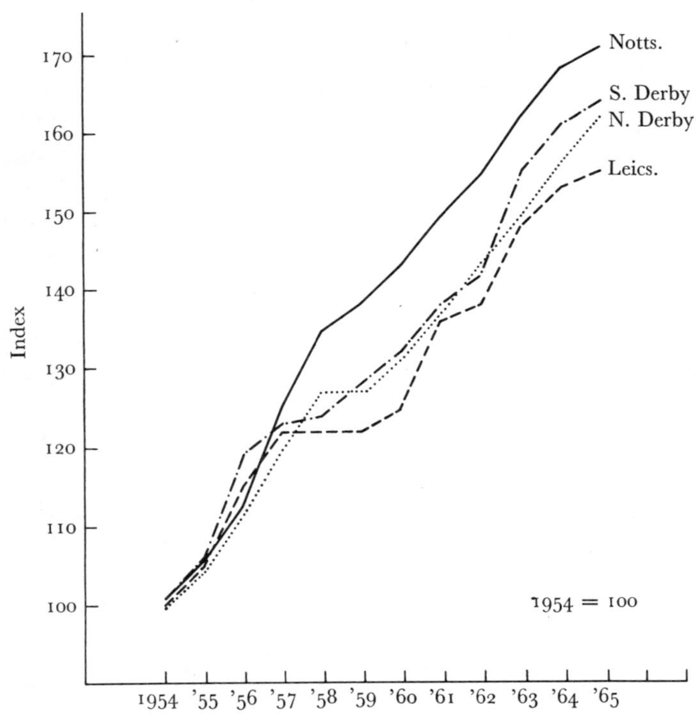

Figure 7.3 Face pieceworkers' earnings per manshift in the East Midlands division
Source: NCB, Financial Returns.

Insofar as power loading was concerned, the East Midlands was the most technologically advanced division of the NCB in the early 1950s. The first power-loading agreement in that division was negotiated in Nottingham in 1952. This was followed by agreements in South Derby, Leicestershire and North Derby.

The physical proximity of the coalfields, as well as their administrative unity in the East Midlands Division, made it easy for wage practices adopted in one wage district to be copied in the others. Thus, before the negotiations of power-loading agreements in Leicestershire and South and North Derby, local officials in those coalfields were already applying a mixture of local rates and those fixed in the 1952 Nottinghamshire agreement. And, when power-loading agreements were eventually negotiated, they tended to follow that of Nottinghamshire.

And as it happened, the Nottinghamshire agreement turned out to be a very poor one from the point of view of wage regulation and, in the knowledge that Nottinghamshire coal had a ready market in the local power stations and the future of the coalfield was relatively well assured, its loopholes were exploited to the full. The most abused clause in this context was that which specified output norms in terms of lineal, rather than square, yardage, i.e. the work norm was fixed in terms of the length of face cut, rather than the amount of coal obtained. This gave rise to all sorts of deviant cutting practices and allowed earnings to rise much faster than if square yardage norms had been agreed. The wage malpractices spread to the other East Midlands coalfields and, because of a traditionally strong element of fair wage comparison between them, a remarkable similarity in face pieceworkers' wage increases in all four coalfields was achieved until 1958 when the central authorities agreed a universal freeze on piece rates. Figure 7.3 shows that the success of the freeze differed quite markedly between the coalfields, however, which accounts for the subsequent disparities in earnings.

The clamp-down on face pieceworkers' earnings in the East Midlands was most vigorously pursued in the Leicestershire coalfield, where miners had most readily accepted method study, followed by North and then South Derby. Central wage regulation was also given more effect in Leicestershire, when in 1963 the NCB got the NUM to agree to change from a lineal to square yardage work measurement following an award by the District Referee.

In South Derbyshire, too, where, through a relaxation of work norms, attempts had been made to keep up with wage movements in Nottinghamshire, increases were increasingly linked to more systematic work-norm estimation, although never with the success achieved in Leicestershire. In North Derbyshire, a particular problem for wage control was posed by the ready market which North Derbyshire coal had in the area. This meant that pit managers were prepared to pay almost any price to the men in order to meet the market, and this helps to account for the relatively high level of wage increase in that particular area.

It would appear from the above account that, while there were initially strong forces of fair wage comparison operating to influence wage movements in the East Midlands coalfields, wage differences did arise which are best explained by the varied approach of the NCB and NUM to wages in the coalfields concerned. The success of the wage control function in the East Midlands Division as a whole, however, was clearly reduced by Nottinghamshire's strong bargaining position and its role as wage leader.

Michael Barratt Brown[3] has suggested that it was the spread of power loading *per se* which influenced the relative movement of face pieceworkers' EMS, rather than the gains in productivity which followed its introduction. In support of this, he used evidence from North Derbyshire which showed that face pieceworkers' EMS grew more rapidly in 1958 and 1962, when the spread of power loading was most rapid but the increase in productivity was not. This was certainly the case, but the chain of reasoning is dubious, for it ignores the central constraint on piecerate revisions in the intermediary period. His suggestion might confirm instead that central wage directives had a significant effect in the East Midlands coalfields, other than Nottinghamshire, over the period.

The Nottinghamshire officials were accustomed to criticism from national headquarters over what, in the light of wage movements elsewhere and the egalitarian objectives of the lower-paid areas, seemed rather high rates of face pieceworkers' wage advance. In view of this, Figure 7.4, which shows the twelve month moving average of EMS and face productivity on power-loading and machine-cut faces in Nottinghamshire between January 1962 and December 1965, is especially interesting. It shows that after 1962 face pieceworkers' EMS on power-loading faces in fact rose more or less in line with productivity. Although one could argue that it was the machines which were responsible for the productivity gains, and in this sense wages should not have risen as fast as they did, productivity does at least provide a rationale for the trend of face pieceworkers' earnings on mechanised coalfaces over the period. Equally interesting, however, is the trend of face pieceworkers' EMS on non-mechanised coal faces, which is virtually identical to that on power-loading faces. This pattern was repeated in every coalfield we examined, and

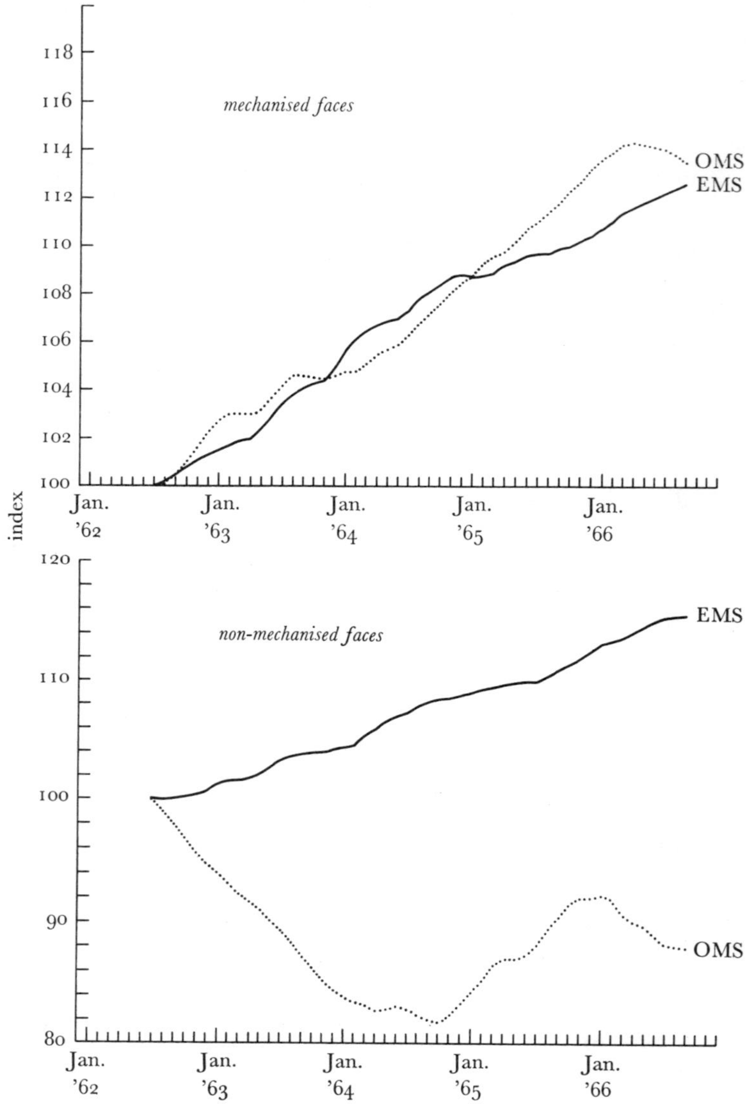

Figure 7.4 Face pieceworkers' EMS and face OMS in Nottinghamshire, monthly (de-seasonalised) 1962–66
Source: NCB, Monthly Statement of Earnings and Productivity.

illustrates how the earnings norms established on power-loading faces were used to regulate wages on conventional faces, despite their declining productivity.

Whereas above average increases in productivity in the mid 1950s helped the already relatively high-paid areas to recoup some of the advantage they had lost during the earlier period, this trend was reversed as power loading spread to the other coalfields. The subsequent contraction of the wage structure was also helped by increasingly effective wage regulating procedures in the Midlands coalfields, which offset some of the more glaring disparities which might have otherwise arisen.

But we are still left to explain why relative movements in face pieceworkers' earnings should have displayed a greater sensitivity to relative economic performance after the mid 1950s. The changed market position of coal in the late 1950s might provide part of the answer.

A move towards wage equalisation would imply that, on average, wages would have to rise faster in the least profitable coalfields. EMS did indeed tend to rise fastest over the period 1950–56 in what, in 1950, were the least profitable areas ($r=0.46$), and many coalfields making a profit in 1950 were making a loss in 1956.

Variations in the rate of change in profitability between the coalfields was related to differences in the rate of change in productivity between them ($r=0.79$). At any fixed output, increases in productivity must be reflected in declining employment and, at any fixed employment, they must be reflected in increased output; it is thus not surprising that coalfields with above-average increases in productivity tended to have below-average increases in employment ($r=-0.51$). But percentage changes in employment and wages were inversely related ($r=-0.48$). This would imply a positive association between percentage increases in wages and productivity. But this was not so ($r=-0.17$). Thus EMS increased fastest where output rose least ($r=-0.57$). Indeed, output and employment declined in many of the relatively low-paying coalfields between 1950 and 1956. Thus the wage structure contraction which occurred in that period did so against the background of a higher rate of marginal pit closure in the least profitable areas.

We should not infer too much from these findings. Pits would have closed faster in areas like Durham anyway, because reserves were being exhausted more quickly. But the evidence does suggest that the wage pattern emerging in the early 1950s was at odds with conventional economic logic. This was unimportant, perhaps, so long as the NCB was forced to keep open otherwise unprofitable collieries, although it might explain why Durham, in particular, kept a tight reign on pieceworkers' EMS over the first decade of nationalisation. But in the next decade costs were important, particularly in the high-cost areas where profits had been adversely affected by previous wage trends. The threat to employment posed by a further substantial contraction of the wage structure not supported by increased productivity must have been evident to both managers and union officials.

The Revised Plan for Coal[4] of 1959 indicated that output would expand in the low-cost pits of Yorkshire and the Midlands, and also in South Wales, which, although a high-cost field, had in steel a ready market for its specialist coals. Output in Kent and Scotland was to stabilise. But in all other fields, and particularly Northumberland and Durham, output was to be reduced.

The Plan was soon followed by a revision of the NCB's pricing policy. It now became undesirable to produce coal at a loss in the high-cost areas. Yet to increase prices generally would have hampered the low-cost areas in their competition with rival fuels. Consequently, the NCB sought to shift the burden of high production costs onto the coalfields generating them. Thus, whereas the price of large coal, which was still in plentiful demand, was raised by a uniform amount in all fields, the price of other coal was raised specifically in the high-cost areas. These changes placed new pressures on the high-cost areas and enhanced the differences in the potential bargaining strength between them and the low-cost areas.

7.5.2 *Daywages*

The scope for adjusting to local situations was clearly greatest in the case of pieceworkers. But the wages of other work groups do not appear to have been wholly immune from local influences. The wage structure of pieceworkers working behind the coalface and of daywagemen contracted more than that of face pieceworkers. But local decisions do seem to have played a part in preserving inter-coalfield league positions. This might suggest that, although achieved by different mechanisms, local manoeuvring on daywages was influenced by considerations similar to those which affected the wages of pieceworkers if not by the actual behaviour of pieceworkers' wages.

There were, and still are, several ways in which daywagemen's EMS could be maintained above the appropriate standard wage rate. First, men could work overtime hours. Second, they could be upgraded. Third, they could be given allowances for working in unpleasant conditions or for doing special jobs. All these could be perfectly legitimate methods of raising pay without abrogating national agreements. But they could equally be means of circumventing national agreements. As we are using highly aggregated data, our figures could also reflect variations in the skill of the labour force and different wage structures between coalfields.

Of these factors, overtime undoubtedly has been the most important single item and the most prone to fluctuation. We shall say more about the role of overtime in wages after 1966 shortly. For the time being, we shall restrict ourselves to the period under consideration. In 1965 overtime shifts worked by surface daywagemen varied from 1.12 a week in Yorkshire, to 0.63 a week in Cannock Chase, and between 1.08 shifts a week in South Derbyshire and 0.56 in Scotland in the case of daywagemen employed behind the coalface.[5] By regressing overtime shifts worked against EMS, we calculate that variations in the number of overtime shifts worked per week explained 44% of the variance of EMS between the coalfields among surface daywagemen, and 38% in the case of elsewhere underground daywagemen.

It is not easy to provide statistical evidence to support the contention that overtime was used purely to enhance the pay of daywagemen, although it has become particularly associated with evasive wage practices. There are several good reasons why overtime might be necessary. It can be a convenient method of overcoming labour shortages and, with the increase in the non-variable cost of employment, it can sometimes prove cheaper than employing additional men. Some productive processes are dependent on flexible working hours and where the 40 hour standard week is operating this flexibility often means overtime working, as in the case of coalface working.

It is most unlikely that any one of the above reasons for overtime working is sufficient to explain its distribution between work groups. Craftsmen were, and still are, expected to work up to 2 or 3 extra shifts in overtime hours on machine maintenance. This partly reflects shortages of some types of craftsmen. But overtime is also worked to offset what might otherwise be adverse consequences for the production process as a result of absenteeism. If it were used for purely economic reasons, such as overcoming labour shortages or production bottlenecks, we might expect these factors to have affected pieceworkers as well as daywagemen and to see a positive association between the number of overtime shifts worked by the two groups. Although very little overtime was actually worked by pieceworkers, we did find a positive association between overtime worked by pieceworkers and daywagemen ($r=0.20$). However, it proved to be statistically insignificant at more than the 10% level.

There was, however, a statistically significant relationship between the level of face pieceworkers' EMS and the number of overtime shifts worked by daywagemen employed at both the pit surface ($r=0.51$)

and behind the coalface ($r=0.72$). So it might have been used deliberately to reduce the wage differential between pieceworkers and daywagemen. And this is a large contributory factor in explaining why, even in 1965, there was a positive and statistically significant relationship between the level of EMS between face pieceworkers and daywagemen working at both the pit surface and elsewhere underground. In this context, it is interesting to find a statistically significant relationship in 1965 between the EMS of daywagemen and wage district productivity ($r=0.56$). While this is not so strong as that between face pieceworkers' EMS and productivity in the wage districts ($r=0.65$), it does tend to suggest that those districts where productivity (and profitability) was highest adopted a more liberal attitude to overtime among daywagemen. Even when we excluded overtime, however, there was still a positive association between pieceworkers' EMS and that of daywagemen, as well as between daywagemen's EMS and productivity. Let us examine some of the other ways in which these might have been brought about.

The 1955 daywage agreement provided that similarly-graded daywagemen should eventually be brought on to the same standard grade rate irrespective of the coalfield in which they worked. Daywagemen whose EMS was above the grade rate were allowed to keep their so-called personal rates until they changed their jobs or until the rates were overtaken by increases in the standard grade wage rates for their particular jobs. The incidence of these personal rates was by no means uniform between the coalfields. There was a higher proportion of personal rate men in the Midlands coalfields, and Nottinghamshire in particular, than elsewhere. Under pressure from the relatively high-paid areas, though, it was decided nationally to raise the wage ceilings periodically. This inevitably delayed the absorption of personal rates into the structure, and explains why, even in 1966, there was a small but significant proportion of daywagemen still receiving them. This must be part of the reason for the difference in EMS levels which persisted between daywagemen in 1966.

But personal rates also persisted because of the reluctance of some areas to abide by the terms of the 1955 Agreement and subsequent revisions. In Nottinghamshire, for example, where daywagemen as a group retained a fairly substantial wage advantage over their counterparts in other districts, one of the Area Industrial Relations Directors stoutly refused to abide by the terms of the agreement for fear of its effect on labour recruitment and wastage. As very little was done about this until 1958, when he was removed from office, the malpractices quickly spread through the Midlands coalfields.

With technological change, we would expect the skill mix of the labour force to change. The number of craftsmen increased from 9% to 16% over the period 1955 to 1966, for example, while the proportion of men in the highest non-craft underground grades increased from 64% to 71%. The highest proportion of craftsmen, who received slightly higher national wage increases, and highly-rated daywagemen was to be found in the more technologically advanced coalfields of the East Midlands. But the 'Black Book' containing job descriptions failed to keep pace with the impact of technology on particular jobs, especially those underground, and in the meantime changes were catered for in a less official manner. The task of transporting supplies is a good example of how local managers dealt with the problem. In 1955, when few pits were mechanised, it consisted mainly of carrying wooden pit props to support the coalface. Once fixed, the props were left to be buried. As technology advanced, steel props were used which, because of their cost, had to be retrieved. At the same time, the transport system was extended underground. Thus, the supplies transporter had to make more journeys, loading two or three times a shift, with heavier equipment. These men gradually became known as salvage workers, and because of the cost of the equipment they were handling, were paid an enhanced task rate although they often received the five-day-week bonus of 20% of base rates applicable to daywagemen, rather than the 16% applicable to pieceworkers. Again, such practices were more common in the East Midlands coalfields than elsewhere. As jobs treated in this manner were classified as underground piecework tasks, such manipulation does not affect our figures for daywage EMS. Nevertheless, it does affect the average EMS of particular coalfields.

We do not suggest that the very substantial increase in the proportion of underground pieceworkers in the labour force, which, over the period 1956–65, ranged from 8 percentage points in South Derbyshire to 2 percentage points in Durham, should be attributed only to the practices described above. But if they could be explained wholly by national policy directives made in the light of technological change, we would expect the percentage of pieceworkers working underground to be related to levels of technology. Of this there was only a slight indication. In part, then, the higher percentage, as well as the above average increase, in the proportion of pieceworkers elsewhere underground in the more profitable coalfields probably reflects a more generous interpretation of the daywage structure. As we have noted already, these anomalies were a major obstacle when it came to revising the structure for pieceworkers working behind the coalface.

7.6 Structural developments after 1966

The piecework system was clearly an obstacle to achieving consistent wage relationships. The gradual abolition of piecework in the late 1960s, with first the NPLA and then the series of national agreements which culminated in the Third Daywage Agreement of 1971, facilitated a more systematic approach to the problem of wage differentials.

The objects of the 1966 and 1971 Agreements were twofold. First, they sought to achieve greater wage uniformity between the previously piecerate tasks which they encompassed. Second, they sought to reduce the wage differentials between the various skills covered by these agreements, and between these and the traditionally daywage tasks paid according to the 1955 Agreement.

If these twin objectives were to be achieved, workers conditioned to the NPLA agreement had to receive a smaller rate of wage increase on average than traditional daywagemen, as did pieceworkers working elsewhere underground in the run-up to the 1971 agreement. At the same time, the highest-paid NPLA workers and elsewhere underground pieceworkers had to receive smaller percentage wage increases than their lower-paid colleagues. Given the very large disparity in earnings between both NPLA workers and pieceworkers working elsewhere underground in the various coalfields, this twin assault on differentials necessarily implied very uneven rates of wage increase between different occupational groups within and between the coalfields.

The NCB had given the NUM an undertaking in 1966 that it would achieve parity in NPLA standard wage rates within five years, i.e. by June 1971. This objective was hampered by the Labour government's incomes policy, under which miners received no wage increase in 1969. After this setback, the wage explosion helped to redress some of the lost increases and a parity of NPLA rates was achieved in December 1971, thus being delayed by only six months.

The NCB had thought in 1966 that NPLA parity could only be achieved in five years if wage rates in Kent and Nottingham remained frozen for that period. This was asking a great deal of the NPLA workers in these areas, many of whom in any case were suffering wage cuts as they moved to NPLA faces. In fact the NPLA operators in the highest-paid coalfields did receive small increases. This was necessary in view of rising retail prices and did not conflict with the parity objectives because of the higher rate of wage inflation generally. Even so, the increases given to the highest-paid ex-pieceworkers were relatively small. Between 1967 and 1970, for example, the highest-paid NPLA workers in the coalfields of Kent, South and North Nottinghamshire received increases of only between 4.4% and 9.5%, compared with 17.4% for NPLA workers in West Wales – one of the lowest-paying areas in 1967. The range of increases given to elsewhere underground pieceworkers was even larger – from 7.8% in Northumberland to 25.3% in Doncaster. NPLA workers received an unweighted average wage increase which was less than half that given to daywagemen working elsewhere underground and at the pit surface, while the unweighted wage increase for elsewhere underground pieceworkers was between that of NPLA workers and surface daywagemen.

There can be little doubt as to the effectiveness of national wages policy in achieving its twin wage structure objectives. Table 7.9, for example, shows that the coefficient of variation of shift earnings fell substantially in the case of miners conditioned to the NPLA and Third Daywage Structure Agreements. Indeed, in both cases the coefficient of variation was smaller in 1974 than for traditionally daywage tasks. Again, turning to Table 7.10, the skill differential was more than halved in every coalfield and the spread of inter-coalfield differentials was reduced to negligible proportions. The differences in skill differentials which remained were due almost entirely to variations in working hours between the coalfields and occupational groups.

We noted in Chapter 6 that overtime had increased quite substantially since the abolition of piecework underground, particularly at the coalface. In its report on earnings in the South Nottinghamshire and West Wales coalfields, the NBPI[6] also noticed this tendency and attributed it, in part, to the inclusion of craftsmen who tend to work more overtime hours than other workers in the face teams. If this were the main factor, we would not expect it to affect the all-industry overtime figure. In fact, overtime rose markedly over the period 1967–70, except at the pit surface, where workers were

Table 7.9. *Coefficient of variation of coalfield EMS 1967–74*

	NPLA	Third daywage structure	Elsewhere underground daywagemen	Surface daywagemen
1967	8.0	6.6	2.2	2.8
1970	4.0	4.1	2.1	2.7
1974	1.4	1.5	1.7	2.1

Table 7.10. *Occupational EMS differentials 1967–74*

	NPLA/Surface daywage 1967	1970	1974	NPLA/Third day-wage structure 1967	1970	1974	Third daywage structure/ Elsewhere under-ground daywagemen 1967	1970	1974
South Notts.	167	155	125	120	108	105	143	136	112
Kent	186	153	126	121	113	107	141	126	106
North Notts.	182	155	125	120	110	107	148	137	110
South Mids.	177	155	131	127	106	106	126	125	112
North West	176	150	127	120	109	105	138	129	109
Doncaster	175	152	123	114	105	105	147	140	112
North Derby	163	155	126	122	109	107	139	134	110
Barnsley	174	153	128	121	112	106	135	129	111
North Yorks.	175	155	128	104	111	106	155	130	110
South Yorks.	171	150	127	116	113	107	142	131	112
Staffs.	171	150	—	114	106	—	139	134	—
South Scotland	170	153	130 (South and North Scotland)	95	103	104 (South and North Scotland)	159	135	110 (South and North Scotland)
North Scotland	167	151		108	103		138	132	
East Wales	154	142	123 (East and West Wales)	112	105	106 (East and West Wales)	134	132	108 (East and West Wales)
West Wales	158	144		113	109		129	124	
Northumberland	176	155	127 (Northumberland, North and South Durham)	104	109	106 (Northumberland, North and South Durham)	153	131	108 (Northumberland, North and South Durham)
North Durham	163	158		102	108		132	126	
South Durham	160	147		113	109		129	124	
Average	170	151	129	113	108	106	140	131	110
Difference (highest–lowest)	32	16	8	32	10	3	33	16	6

Source: NCB, Financial Returns.

unaffected by the changeover from piecerate to daywage payment. This suggests that the increase in overtime at the face was more than a statistical illusion created by the inclusion of craftsmen in the power-loading teams. So, too, does evidence[7] on overtime at the face in the various coalfields.

Overtime premium shifts as a percentage of total shifts worked at the coalface in 1967 varied from 2.8% in Northumberland to 8.3% in North Yorkshire. In 1970 they varied from 3.7% in North Durham to 14.6% in the South Midlands. The increases in overtime shifts varied from 0.1 percentage point in North Durham to 6.6 in the South Midlands. Increases in the proportion of overtime shifts worked were unrelated to the increased number of faces operating under the agreement. In Durham and Scotland, for example, where all faces came under the NPLA in June 1966, the overtime percentage was lower in 1967, and had risen much less rapidly by 1970, than in the East Midlands coalfields and many others, where, even in 1970, not all faces were on the agreement. One plausible explanation of the increase in overtime working might be that pit managers were offering overtime as a wage incentive in the absence of piecework. We shall develop this point in Chapter 9. Suffice it to say here that pit managers were finding it increasingly necessary after the NPLA to rely on overtime to meet production targets. It is also interesting to note that, although it was by no means very strong, there was an inverse relationship between the percentage increase in overtime worked at the face and the increase in basic shift rates between 1967 and 1970. And a comparison of overtime in the six coalfields where basic shift rates rose fastest and those where they rose least, revealed that it increased by 2.9 percentage points in the former, but by 4.8 percentage points in the latter. It would seem, therefore, that overtime was being used to compensate for below-average increases in standard shift rates.

7.7 Summary and conclusions

Although central wage agreements became increasingly effective in the coalfields in the long term, their impact varied according to local circumstances. National decisions were most easily put into effect in the more cost-conscious coalfields, particularly in the case of faceworkers, after the slump in coal demand in the 1950s. Our analysis shows that the pace of technological change and productivity improvement, the style of wage agreements and the geographical proximity of the coalfields, were all important influences in determining pieceworkers' wage movements before the revisions of their wage structures, and that they helped to mitigate the impact of national agreements. These influences were also seen to spin off in the wages of other mineworkers. With the revision of the pieceworkers' wage structures local managements seem to have looked to overtime to provide some flexibility in the wage system.

8. EXTERNAL WAGE RELATIVITIES

As yet we have said little, except by implication, about the relationship between wage movements in mining and other sectors of the economy. This is not because we think pay movements outside mining and external pay comparisons are unimportant for a study of coalmining wages policy and behaviour – they certainly are. On the contrary, it is because the subject is sufficiently important and complicated to warrant separate examination. It is also because the significance of external wage movements over much of the period under consideration is best appreciated in the context of internal wages policy. Indeed, the inter-relationship between internal and external wage movements is of particular interest to the coalmining case, and the subject is thus best treated as a unified whole.

The actual behaviour of miners' wages in relation to those of other groups has been a matter of considerable controversy. This is partly due to differences in the accounting procedures used to compile earnings in coalmining and those adopted for the D of E's earnings enquiries. We commence this chapter with an account of these differences and some of their consequences. In the second part, we concentrate on the inter-relationship between internal and external wage movements. Our purpose there is to illustrate how internal wages policy and wage movements in mining have been affected by wage movements elsewhere, and also how internal policy has been instrumental in fashioning external wage relationships. Last, we consider some implications of our findings for future wages policy.

8.1 The data

Wage comparisons are usually based on average weekly or hourly earnings. The data most frequently employed for this purpose are those collected by the D of E from its six monthly earnings enquiries, which relate to one week in April and October each year.[1] However, the D of E (and its predecessors) has never included coalmining in its enquiries: instead it has relied on earnings data supplied independently by the NCB. But the NCB's data, which also refer to one week in April and October, reflect that organisation's wage accounting conventions, which differ in some quite important respects from those which other employers are asked to follow for the purpose of the D of E's enquiries.

The first set of problems in comparing NCB earnings data with those of the earnings enquiries stems from differences in coverage. The NCB's figures include foreman grades, such as underofficials and weekly paid industrial staff, while the D of E earnings enquiries specifically exclude them. Next, the NCB's definition of earnings is taken to include payments for sickness and annual and statutory holiday payments, which are excluded in the data supplied through the D of E enquiries. On both these counts, NCB accounting procedures inflate miners' earnings relative to the manual workers' earnings reported by the D of E. And as payments for sickness and holidays have become increasingly important in the miners' earnings, the effect of including them becomes greater with the passage of time.

The second group of problems stem from the format in which NCB data are compiled. While the NCB provides details of average weekly earnings of adult males over 21 years of age for both April and October on the lines outlined above, it does not provide a statistic of hourly earnings, or of the hours worked. Rather, it follows its own conventions and reports earnings on a 'manshift worked' basis. As we know the hours equivalent of a shift at the pit surface and underground from national agreements, we might, with information about the proportion of shifts worked at each place of work, reasonably derive a statistic of hourly earnings from the EMS figure. But the EMS figure which the NCB publishes for April and October refers to all mineworkers which, in addition to foremen grades, includes juveniles. It is impossible, therefore, to calculate a statistic of average hourly earnings for *adult* mineworkers directly from the published EMS data.[2] These problems have not prevented people from making pay comparisons between miners and workers in other industries. But because these problems have often been ignored, or assumed away, such comparisons have led to considerable controversy about what has actually happened to miners' external wage relativity[3] since nationalisation.

While we cannot hope to solve all these difficulties, an awareness of them is essential if we are to provide a realistic account of the trend of miners' relative pay.

8.2 The miners' external pay relativity

As the D of E discontinued its April earnings enquiry in 1970, we shall base our analysis of pay comparisons on the figures collected for October between 1948 and 1974.

Table 8.1. *Average hourly & weekly earnings of adult males in some industries as a proportion of the all-industry average 1948–74*

	1948	1956	1960	1966	1970	1974
Coal						
Hourly 1[a]	144	149	139	130	116	141
Hourly 2[b]	139	141	130	119	102	121
Weekly 1[a]	123	129	112	109	100	120
Weekly 2[b]	122	122	105	99	89	103
Food, drink & tobacco						
Hourly	90	90	91	95	98	95
Weekly	93	92	93	97	99	99
Chemicals						
Hourly	100	101	105	107	107	108
Weekly	100	101	103	105	105	105
Metal manufacture						
Hourly	113	113	113	108	108	107
Weekly	114	113	110	106	107	106
Engineering						
Hourly	107	106	106	103	102	100
Weekly	107	107	105	101	100	98
Shipbuilding						
Hourly	103	102	102	105	106	107
Weekly	103	103	99	105	106	104
Vehicles						
Hourly	115	110	125	121	125	116
Weekly	113	106	117	108	116	108
Textiles						
Hourly	93	92	92	92	92	93
Weekly	94	92	92	91	90	90
Paper, printing & publishing						
Hourly	110	118	118	119	121	116
Weekly	106	115	117	118	120	113
All manufacturing						
Hourly	105	104	106	105	105	104
Weekly	194	103	104	102	103	101
Construction						
Hourly	94	96	92	96	92	97
Weekly	95	99	96	101	96	100
Gas, elect. & water						
Hourly	95	93	96	99	96	101
Weekly	95	92	96	94	93	98
Transport & communications						
Hourly	92	91	93	94	98	98
Weekly	95	96	99	103	106	107

Table 8.1. (cont.)

All industries						
(£s) Hourly	0.148	0.245	0.303	0.441	0.614	1.078
Weekly	0.895	11.895	14.535	20.305	28.050	48.630

[a]including holidays and sickness pay
[b]excluding holidays and sickness pay
Source: *Department of Employment Gazette*

The first mining to 'all-industry' average weekly earnings relativity given in Table 8.1 is based on the average weekly earnings statistic for adult mineworkers in October each year supplied by the NCB to the D of E. It thus includes payments for annual holidays, rest days and sickness benefit, details of which were given in Chapter 6. The second weekly earnings relativity is based on a figure of miners' earnings which excludes these supplements. We could have excluded foremen grades after 1955. But, as it was impossible to do so before then, we have included them throughout the period. The main effect is to increase slightly the absolute level of miners' earnings, rather than to affect their relative trend.

The hourly earnings figures for mining have been derived in the following manner. First, we divided the October figures of average weekly earnings for *all* mineworkers by the October EMS statistic of that same group to derive the number of shifts worked. We then multiplied this figure by the *weighted average* of the number of hours in a shift worked at the surface and underground,[4] to get the hours equivalent of a shift. On the assumption, which we feel to be realistic, that this figure is not substantially different from that worked by adult mineworkers, we then calculated average hourly earnings by dividing the average weekly earnings of *adult mineworkers* by the derived figure of hours worked. Our first estimate of hourly earnings thus includes the imputed value of payments for holidays, rest days and sickness, while the second figure excludes these items.

Miners work fewer hours than male workers in other industries. This gives them a more favourable relativity when it is measured in hourly, as opposed to weekly, earnings. At nationalisation, miners as a whole were the highest paid of the major industrial groups in terms of both hourly and weekly earnings, irrespective of which series one chooses for comparative purposes. They maintained, and slightly improved upon, this position until the mid 1950s. Between 1956 and 1960, their relativity in both hourly and, particularly, weekly terms declined substantially, due to the sharp reduction in hours worked in the late 1950s. The decline in the miners' relative pay position continued until the early 1970s. Our figures suggest that between

1960 and 1966 the relativity in hourly terms fell by between 9 and 11 percentage points (depending on which estimate one chooses) and between 1966 and 1970 by between 14 and 17 percentage points. Then, following two national strikes, in 1972 and in 1974, it was restored to a position approximating that of the late 1950s.

The actual relative pay position of miners in any year does vary according to which estimate of hourly and weekly earnings one chooses for comparative purposes. Moreover, the difference becomes greater the more recent is the period. If one takes the hourly earnings relativity, for example, a discrepancy between our two series of 5 percentage points in 1948 increased to 8 percentage points in 1956 and 20 percentage points by 1974. Over the same period, the discrepancy between the two weekly earnings series rose from 1 percentage point to 17. By 1974, the difference between the estimates was sufficient to put miners' hourly earnings either 41% or only 21% above the all-industry average, and their weekly earnings either 20% above, or 3% above the all-industry average.

The degree of variation in the miners' earnings relativity in either hourly or weekly terms over the period 1948–74 was unrivalled by any other industry. The largest variation in earnings relativity outside coalmining was in vehicles, and paper, printing and publishing. In the former, hourly earnings varied by 15 percentage points and weekly earnings by 11 percentage points from the all industry average. In paper, printing and publishing, the variations were in the order of 11 and 14 percentage points respectively. These figures compare with a variation of nearly 30 percentage points in coalmining. Clearly, then, miners' wages have displayed an unusually high degree of sensitivity to the particular circumstances confronting their industry.

Before we proceed further, it is interesting to compare our results, particularly in the late and early 1960s, with those of other commentators on coalmining wage movements over that period. It will be remembered, for example, that Turner[5] suggested that, from the end of the Second World War until the mid 1960s, miners' earnings had kept pace with wages in other industries. As a general statement, this might be true, but the more interesting question might have been – why, having kept pace with wages elsewhere until the mid 1950s, did they fall behind thereafter? This also puts a slightly different complexion on what Turner[6] identified as a wage drift of 1% per annum up to the mid 1960s. This, he suggested, was the attempt by miners to keep their wages moving at the same pace as those in other industries. While such an attempt might indeed explain such wage drift as there was, the drift itself certainly did not achieve that effect after 1956 if – like Turner – we take miners as a whole. Again we must take issue with Michael Barratt Brown,[7] who expressed surprise that, after a fall in the miners' relativity in the late 1950s, miners' hourly earnings did not fall by more than 1 percentage point between then and 1966. On our estimates they did – by at least 9 percentage points. We would not disagree with his analysis that the increase in overtime pay helped to sustain miners' earnings at a higher level than would otherwise have been the case, or that this might have represented an attempt to offset some of the lost relativity in the period. But we cannot accept that increased overtime maintained the relativity of miners as a whole between 1960 and 1966. To accept the argument of fair wage comparisons, then, is to accept, not unreasonably, that it can be associated with any trend in the wage relativity of a particular work group. This is amply illustrated if we examine the external relativity of surface daywagemen, pieceworkers and taskworkers working at the face, and those working elsewhere underground, over the same period. This we can do with the help of Figure 8.1, where we have drawn a few lines to guide the reader through the more interesting periods.

After falling briefly behind workers in manufacturing industries during the early postwar wage freeze, surface daywagemen increased their external relativity over manufacturing workers and by 1956 it had been restored to the level of 1948. The relativity of pieceworkers showed considerable fluctuation over this period, but by 1955 was below the level of 1948. Thus the reduced internal wage differential between face pieceworkers and surface daywagemen over this period was reflected in a slightly *reduced* external relativity for pieceworkers and, helped by a very small element of wage drift, a slight *increase* in the external relativity for daywagemen.

Between 1956 and 1960, the external relativity of daywagemen fell by 10 percentage points. Helped by transfers of workers to power-loading faces and the even higher wage rates under power-loading agreements, the pieceworker relativity actually increased until 1958, after which it, too, fell substantially – by 19 percentage points. During this period, then, internal wage structure contraction occurred against the background of a falling external relativity for all mineworkers.

Some of this relativity was regained in the early 1960s, but it was lost again by 1965. With the introduction of the NPLA, the decline in the external relativity of faceworkers accelerated until 1970. During this period, though, the external relativity of daywagemen was maintained at approximately the level of 1965 but below its peak of 1956.

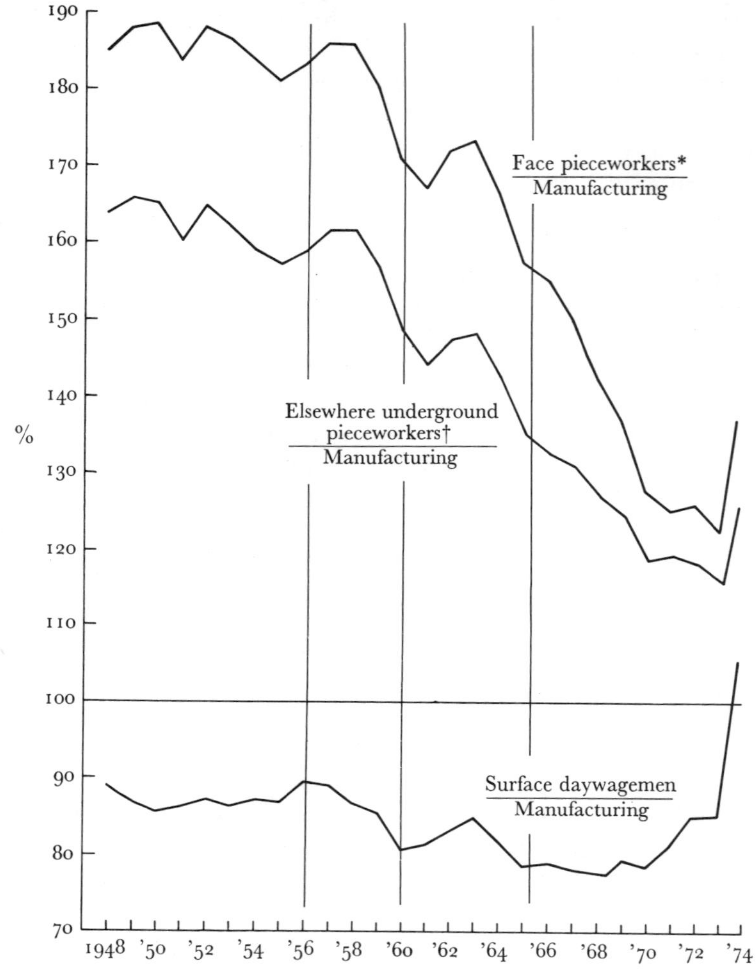

Figure 8.1 Average hourly earnings relativities between manufacturing males, and coalmining pieceworkers and daywagemen 1948–74

* After 1966 earnings are primarily for faceworkers conditioned by the NPLA.

† After 1971 earnings are primarily for workers conditioned by the Third Daywage Agreement.

Source: NCB, Financial Returns; D of E *Gazette*.

These trends, of course, do hide quite large differences in the miners' external relativity in the various coalfields – some idea of which is given in Table 8.2.

In view of the larger internal wage differential of faceworkers compared with daywagemen, it is not surprising that their external relativity should show the largest regional variation. However, those relativities were also affected by differences in manufacturing earn-

Table 8.2. *Average hourly earnings relativities in the regions between coalface and surface miners and manufacturing males 1960–74*

	1960		1965		1970		1974	
	FP/M	SD/M	FP/M	SD/M	FP/M	SD/M	FP/M	SD/M
Including holiday and sickness pay								
South East	195	83	177	81	135	82	154	110
E. Midlands	208	90	192	89	152	88	169	121
Yorkshire	193	89	181	89	148	91	168	117
North West	188	92	187	90	139	86	161	113
W. Midlands	160	77	157	76	128	77	151	104
Northern	170	85	152	83	130	81	158	108
Scotland	178	86	164	86	134	84	159	110
Wales	144	78	140	77	126	80	151	110
Average	180	85	169	84	137	84	156	112
Range	64	13	52	13	26	14	18	17
Excluding holiday and sickness pay								
South East	186	78	165	73	123	72	133	96
E. Midlands	197	83	177	80	138	78	144	103
Yorkshire	183	83	166	79	133	79	142	99
North West	179	85	173	80	127	76	142	97
W. Midlands	152	71	145	68	117	68	131	91
Northern	162	75	140	73	118	71	132	92
Scotland	170	79	151	75	120	71	133	95
Wales	136	73	129	69	111	70	136	94
Average	171	78	156	75	123	73	137	96

FP = Face pieceworkers and NPLA
SD = Surface daywagemen
M = Manufacturing males
Note: With the exception of 1960, manufacturing earnings are those given in the October Earnings Enquiry reported in the *Department of Employment Gazette*.

Coalmining hourly earnings refer to all mineworkers in the accounting year. They are thus centred on June in 1960 and October thereafter. The hourly earnings figures were calculated by dividing shift earnings by the appropriate hours in a shift at each place of work.

ings, which did not always reflect the internal ranking of the coalfields. For example, Scottish faceworkers were relatively low paid by coalmining standards but relatively highly paid by manufacturing standards, whereas South Wales faceworkers had a small external relativity because of comparatively high wages in the local steel industry. East Midlands miners were relatively highly paid by coalmining standards and even more highly paid by standards in manufacturing in that region. Nevertheless, if internal wages policy in mining succeeded in contracting the faceworkers' wage structure it would tend, so long as manufacturing earnings rose at a fairly even pace in all regions, to reduce the degree of external wage dispersion.

The degree of wage dispersion among surface workers was fairly small by 1960, and more so than in manufacturing. Thus, differences in their external wage relativity were determined mainly by differences in manufacturing hourly earnings.

Perhaps the most interesting aspect of this table is the maintenance of the surface daywage relativities in most coalfields over the period 1960–65 and, to an extent, over the period 1965–70, especially if one takes the figures which include holiday and sickness pay, despite what should have been varying rates of wage increase among daywagemen. This could have resulted in one of two ways – either the rate of increase in manufacturing mirrored that of the coalfields, or the pattern of wage increase in the coalfields adapted to the pattern in manufacturing. The following table, which gives the percentage increase in hourly earnings in coal and manufacturing, helps us to answer this question.

With the exception of Scotland, there was a difference of only 1 percentage point in the rate of increase in daywage hourly earnings between 1960 and 1965 and, with the exception of Scotland again, a difference of only 1 percentage point in increases in manufacturing hourly earnings over the same period. The pattern of mining earnings increases should have been rather more varied than they were, if wages policy had been wholly effective in regulating them, so our figures suggest that external wage comparisons were an important influence in determining local wage increases to daywagemen. The

Table 8.3. *Percentage change in average hourly earnings of adult males in manufacturing and of mining faceworkers and surface workers (including holiday and sickness pay) 1960–74*

	1960–65			1965–70			1970–74		
Region	FP	SD	M	FP	SD	M	FP	SD	M
S. East	28	36	40	11	43	46	97	138	72
E. Midlands	28	37	39	16	45	46	92	138	73
Yorkshire	30	38	39	19	47	45	104	132	80
N. West	40	37	41	13	45	53	100	130	74
W. Midlands	37	37	39	20	47	46	97	125	67
Northern	26	38	41	27	44	49	112	132	74
Scotland	31	41	42	25	48	53	110	133	76
Wales	35	36	39	29	49	44	101	134	68

Source: Manufacturing: Department of Employment, *Gazette*
Coalmining: NCB, Financial Returns.

pattern was also influenced by the relaxation of the ceilings policy under external wage pressures. The higher manufacturing wage increase in Scotland might also help to explain why surface miners there got slightly larger increases than miners in other coalfields.

A similar set of influences can be seen to have worked on faceworkers' earnings, but with less effect. The much faster rate of wage increase in Lancashire and Cheshire, for example, kept face earnings there up to the increase in manufacturing earnings, while the slightly slower growth of face earnings in the West Midlands and Wales was not substantially less than those of manufacturing workers. The much tighter control over face earnings in the other coalfields, however, did reduce the faceworkers' external relativity.

We can see similar influences working on daywages in the period 1965 to 1970, but with much less effect, due to a rather more varied rate of increase in manufacturing earnings. The pattern of daywagemen's increases was rather more consistent with the objects of policy and, as we have illustrated, the internal structure contracted during this period. Even so, the actual rate of increase overall was still being influenced by such things as overtime working. In the coalfields of Yorkshire, the West Midlands and South Wales, an average rate of wage increase of between 45% and 47% was sufficient to maintain the daywagemen's external relativity. In the other coalfields, hourly earnings rose by less than this amount and the external relativity fell, if by varying amounts, due to the rather different rates of increase in manufacturing earnings between the coalfields. Again, a rather higher rate of increase in manufacturing earnings in Scotland may have been an important factor in daywage hourly earnings rising by an above average percentage in that coalfield.

No such relationship between faceworkers' earnings and manufacturing earnings existed in the period 1965–70, and not even in the lowest-paid coalfield did wage increases to faceworkers, conditioned by the NPLA, rise by much more than half the increase in manufacturing rates. The result was a drastic reduction in the external relativity of faceworkers over the period in every coalfield, although that reduction varied from 14 percentage points in South Wales to 45 percentage points in Lancashire and Cheshire.

This pattern of external differential erosion of wages considerably reduced differences in the external relativity between the coalfields. In 1965, there was a discrepancy of 52 percentage points in the external relativity between faceworkers and adult males in manufacturing. By 1970 this had reduced to 26 percentage points. Over the same period, the wage relativity between faceworkers and manufacturing males fell from 169 to 137%.

Again, though, external pay relationships were in no case maintained if holiday and sickness pay are excluded from our calculations. Thus, to the extent that an attempt to keep up with pay elsewhere may have been important in local bargaining, its success was, to some extent, illusory.

8.3 Net real income

The full significance of these earnings trends is best appreciated when one takes taxes and price increases into account. This we do in Figure 8.2, which plots the course of net real income (at 1948 prices)[8] for faceworkers and surface workers against that of adult males in

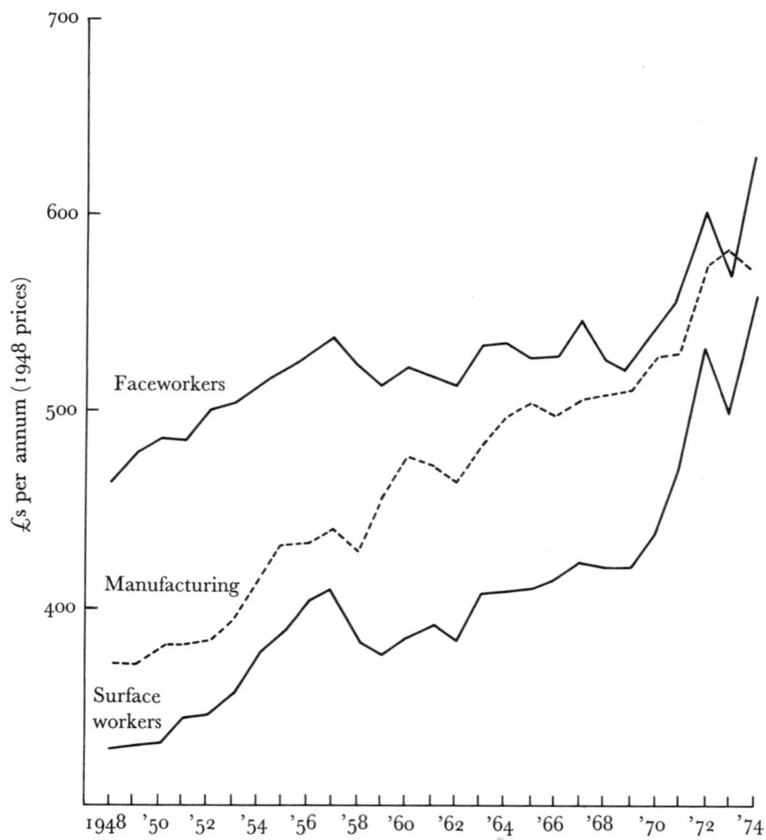

Figure 8.2 Net real income of coalmining faceworkers, surface workers, and manufacturing males 1948–74
Source: NCB, *Report and Accounts*; D of E *Gazette*.

manufacturing. The decline in the rate of wage increase in 1958 was sufficient to produce a sharp fall in the real living standards of all mineworkers, from which neither surface workers nor faceworkers recovered until 1963. Between then and 1970, surface workers had a small increase in real income, but obviously one which was substantially less than that enjoyed by manufacturing males. Apart from a small, temporary, increase in net real income in 1968, faceworkers, on the other hand, found themselves with a net real income in 1970 which was no greater than that enjoyed in 1957.

Discrepancies in the rate of real income growth were even more marked for faceworkers in different coalfields. Between 1967 and 1970, for example, faceworkers in Scotland had an increase in real income of 2.3%, whereas that of faceworkers in North Nottinghamshire *fell* by 1.8%. These figures compare with an increase of 4% in manufacturing over the same period.

The trends described above were, of course, one of the most important factors contributing to the increasingly belligerent attitude of the miners and NUM in the late 1960s. They were, nevertheless, a consequence of an internal wages policy which succeeded in achieving greater wage uniformity between the highest- and lowest-paid groups, while at the same time attempting to depress the external relativity of the lowest paid.

Much of the hostility in the NUM came from the higher-paid coalfields. The squeeze on faceworkers' earnings affected all coalfields, however, and became a common source of grievance and thus a generally unifying influence on the NUM. Furthermore, as the scope for local bargaining was now mainly restricted to overtime working, the only way round the problem was to push as hard as possible in national negotiations. All that was needed now was a change in coal's market environment to permit the NUM to give effect to the increased internal unity given by the wage reforms.

8.4 Bargaining strategy

Of the economic factors important in determining the miners' bargaining power over the period 1949–70, it was not so much the relationship between year-to-year coal production and consumption over the period which emerged as a significant variable in influencing the rate of change of coal EMS as the level of coal stocks.

Thus where

$$g_c = \frac{E_{c1} - E_{c0}}{E_{c0}} = \text{annual percentage change in mining EMS}$$

$$g_m = \frac{E_{m1} - E_{m0}}{E_{m0}} = \text{annual percentage change in manufacturing average hourly earnings}$$

$$S_t = \text{level of coal stocks in million tons,}$$

we have for annual changes between 1949 and 1970[9]

$$g_c = 1.463 + 0.892\, g_m - 0.092\, S_t$$
$$(1.34) \quad (0.134) \qquad (0.040)$$
$$n = 21 \quad DW = 1.80 \quad r^2 = 0.73$$

Each of the coefficients is statistically significant and of the expected sign. The coefficient for manufacturing earnings is positive and less than 1, indicating that over the period increases in manufacturing earnings were associated with smaller earnings increases in mining. This is consistent with our previous analysis. The stock coefficient tells us that for every 1 million ton increase in stocks, there was a downward pressure on EMS equivalent to 0.1 on the percentage rate of change in the miners' average shift rate. In short, it would appear to be the current level of stocks rather than the current addition in stocks[10] which has been influential in determining the size of the percentage change in miners' EMS.

The importance of the industry's stock position in determining the ability of the NCB and the government to resist wage demands was not lost on the NUM. Thus, during the late 1960s – before the increase in oil prices, and before the national strikes of 1972 and 1974 – the miners and the NUM took steps, through overtime bans, to reduce stock levels and thereby increase their bargaining power. This action put the NUM in a particularly strong position by the early 1970s, when first Libya, and then other oil nations, decided on a sharp increase in prices. Steps were taken to slow down the rate of pit closure in 1970–71 and the NCB embarked on an intensive recruitment campaign. In 1971, Lord Robens saw the industry being required to increase production to 150 million tons in the next five years.

These events, which left room for an increase in oil prices and the promise of an increased demand for mining labour, further strengthened the NUM's hand in wage negotiations and removed some of the fears of members in high-cost areas about the consequences of substantially raising the miners' relative pay position. Thus, whereas the NUM's National Executive Committee narrowly failed to get the necessary 66% support for a national strike in 1970, it succeeded in getting over 86% in favour of a national strike in 1972 and a similar mandate in 1974.

The problem of finding a solution to the NUM's grievance was multiplied by the fact that in both 1972 and 1974 the then Conservative Government was attempting to reduce the general rate of wage inflation. In 1972 the government was operating its policy known as '*n* – *1*' which, by the time the miners came to negotiate, implied a settlement in the region of 8%, while in 1974 it was operating a fully fledged incomes policy seeking to restrain earnings increases to around the 10% mark.

While the reports of the Enquiry Boards, established to find a solution to the NUM's grievances, contained many points of interest, it is their treatment of the relativities issue upon which we shall focus attention here.

The NUM's pay claims in the late 1960s and early 1970s were based firmly on the ground that the miners' pay relativity had been eroded and it was on this ground, and the increasingly key position of mining in the UK energy market, that the NUM was afforded special treatment under the pay policies of the period. But the NUM was also arguing that any special treatment should favour the lowest-paid workers. Thus, at one and the same time it was asking that the internal wage structure be contracted further, and that the wage relativity of mineworkers, as a whole, be raised. If the miners' relativity was to rise on average; such a policy clearly implied that the relativity of surface workers would rise more than that of faceworkers, despite the former having fallen less far behind workers in manufacturing. And if the pay package was to raise the external relativity of faceworkers, it would also have to give surface workers an increase which raised their wages substantially above the highest relativity of earlier years. There was a danger, therefore, of any solution falling between two stools. First, as surface workers had more counterparts in other industries than faceworkers, a wage increase which substantially raised their external relativity could spark off competing wage claims in other sectors, which might further erode the faceworkers' relativity Second, a pay package which did not substantially raise the relativity of daywagemen would, in any case, leave the faceworkers' grievances unanswered.

8.5 The Wilberforce Enquiry, 1972

The Wilberforce Enquiry[11], established to find a solution to the miners' strike of 1972, acquiesced in the NUM's demand that the pay of surface workers be raised by a large percentage than that of faceworkers and other workers underground.

The Wilberforce recommendations implied an increase in the

average mining wage level of 20%. Of this, approximately 8 percentage points[12] was given in lieu of the annual wage round anticipated if the government was successful in its policy of wage de-escalation at that time, and the remainder – some 11 percentage points – was calculated to raise the miners' wage relativity to some 10 or 12% above the level of 1971. This would have implied a relativity of 112–113%, if one bases the calculation on the weekly earnings figures of October 1971.

It is possible that the Wilberforce Committee miscalculated the wage increase necessary to achieve and maintain that relativity, and that the government was unduly optimistic about the impact of its own policies on wage increases in other sectors. In any event, between October 1971 and October 1972 manufacturing earnings increased at approximately twice the rate anticipated by Wilberforce, most of the escalation having occurred after the miners' settlement. Thus, by October 1972, miners' average weekly earnings were only 5.6% above those of adult males in manufacturing, despite miners' earnings having risen by 20% over the year.

More important though, whereas by October 1972 the settlement made in March had increased the external relativity of surface workers by 4 percentage points compared with the previous year, the external relativity of faceworkers was no more than one percentage point higher in October 1972 than it had been in 1971. The government implemented a wage freeze in November 1972, and the miners were one of the first groups to settle (in April 1973) under the post-freeze £1+4% policy. By October 1973 the miners' wage relativity was only some 2.3% above that of manufacturing males, or some 10% below that calculated as desirable by Wilberforce. In short, the miners' wage relativity had almost returned to the position of 1971.

Again, though, the relativity of surface workers was maintained during the incomes policy, but the latter served to reduce the faceworkers' relativity below the level of October 1971. In short, the industry's faceworkers had gained no permanent increase in their external relativity. This was probably an important factor in the NUM securing another mandate for a national strike in 1974, when once again it found itself in a head-on collision with the government.

8.6 The pay board, 1974

In considering the miners' case for special treatment in March 1974, the Pay Board[13] was acutely aware of the danger of repercussive pay claims in other sectors. In its opinion, based on evidence given by employers and unions in non-mining sectors, the main threat was posed by the NUM's demand that an upward wage adjustment be combined with a larger percentage increase in the pay of surface workers than of underground miners, so as to compress the internal wage structure still further. This consideration, together with the NCB's evidence that acute manpower shortages existed in underground occupations, convinced the Pay Board of the need to bias an award on the grounds of exceptional treatment in favour of underground mining grades, and that surface workers who had worked underground should receive a larger percentage wage increase than those who had no underground experience.

To quote:[14] 'Our recommendations have dealt with the underground workers because we judge that it is their pay relative to that of all other workers which needs to be changed. We are conscious that in doing so we have altered the internal relativities between underground and surface workers.'

The report went on to say: 'No enduring change in external relativities can be made unless it commands general assent. Without it the change will be quickly eroded. Unless such general assent is forthcoming for an adjustment of the whole of a wage structure, which would be rare and which we judge is not true in this case, a change in the external relativities of some part of a wage structure which is judged to be generally acceptable must imply a change in internal relativities. In this case internal change is a probable consequence of external change.'

The Pay Board's recommendations, which implied a reversal of internal wages policy in coalmining, were only partly acceptable to the NUM. In its subsequent negotiations with the government, the NUM did agree to non-faceworkers *underground* receiving the largest percentage increase in standard wage rates of any group. The NUM claim on behalf of faceworkers (which was slightly less than the amount the Pay Board recommended be given) was met in full. But *surface* workers received a *larger* percentage increase in standard rates than faceworkers.

The impact of the 1974 settlement on external relativities was substantial, particularly in the case of surface workers. Between October 1973 and October 1974, the pay relativity of surface workers rose from 85 to 105%, this latter figure being over 16 percentage points above their previous highest relativity of 89% in 1956. The external relativity of faceworkers also rose substantially, by 15 percentage points – but their external relativity was back to its 1969 level. In effect, then, the attempt to improve the external wage relativity of faceworkers and non-faceworkers underground, in the face of opposition in the NUM to an increase in the internal wage differential between faceworkers and surface workers, did result in an upward movement of the *whole* wage structure.

The significance of this award is seen even more clearly if we refer back to Table 8.2, which gives the 1974 relativity of faceworkers and surface workers in the various regions. The relativity of surface workers in 1974 was at the least 20 percentage points above that of 1960 (South Wales) and at most 29 percentage points above (East Midlands), these differences being accounted for by overtime working. The relativity of faceworkers, compared with 1960, varied from 7 percentage points *above* (in South Wales) that of 1960, to 39 points *below* (in the South East).

The 1974 settlement in mining, which raised average earnings by approximately 32%, was followed by a general escalation of wages in the economy as a whole, although the role of the settlement in this is by no means clear. The immediate cause of that escalation was the triggering in May of the threshold payments admissible under the Conservative government's incomes policy, which the new Labour administration inherited. And, as miners participated in those payments, which by October had raised manufacturing earnings by about 11%, they had little effect on the miners' relativity. In July 1974, the government abandoned statutory pay restraint and embarked on the Social Contract. The new policy marked the beginning of a further wages escalation, although whether this exceeded the terms of the 'Contract' is debatable. When miners came to settle again in March 1975, their earnings were approximately 12% above those in manufacturing, as compared with 18% above in October.[15]

With wages in the economy rising faster than in the previous year, the government, having learnt its lesson from the failure of previous administrations to maintain an increased relativity in mining, allowed the NUM to participate in the general escalation – so restoring some of the lost relativity – before seeking an agreement on wage restraint with the TUC. The NUM settled for a further 19% (excluding threshold consolidations) and the miners' relativity moved to 32%.[16] Again, that relativity was gradually eroded as other groups settled. By August, when the government and the TUC agreed to restrict pay increases to £6 a week, miners' average earnings were approximately 20% above those in manufacturing. Although the flat-rate policy is generally regarded as having been successful in anti-wage-inflation terms, it inevitably had the effect of reducing the mining industry's external earnings relativity: by July, 1976, the miners' earnings advantage over manufacturing males had fallen to 13%. That relativity was eroded still further under the $4\frac{1}{2}\%$ pay policy. That erosion was attributable to a fairly substantial earnings drift in manufacturing. Thus, whereas average weekly earnings in mining increased by 5% between July 1976 and July 1977, average weekly earnings in manufacturing increased by 10%. Once again, therefore, incomes policy operated to the detriment of the miners. So long as we have incomes policies this problem will only resolve itself when the miners have adequate proof that it will ensure equal treatment between the private and public sectors. This, however, might only be achieved with a more flexible wage system in the mining industry – a point to which we shall return later.

8.7 Summary and conclusions

Despite the contraction in the coalmining industry's wage structure in the longer term, which was bound to alter the external earnings relativity of some groups of mineworkers, earnings movements in the rest of the economy have exerted a powerful influence on wages policy and wage movements in the mining industry. With coal in short supply during the first decade of nationalisation, wage movements in the rest of the economy set the floor to both national and local wage determinations in mining, thus preventing more than short-term reductions in the mining industry's internal wage structure.

The slump in coal demand and increase in stocks in the late 1950s temporarily weakened the power of external wage comparison. The NUM became increasingly concerned about employment prospects, particularly in the marginal coalfields, and it was with the latter in mind that national wage increases were agreed. If national wage agreements had been wholly effective in the place of work, the external relativity of both pieceworkers and daywagemen would have continued to fall after the sudden reductions in the late 1950s, caused by the sharp cut-back in overtime working and restraint shown in national negotiations – particularly in the higher-paying coalfields. But further decline in the daywage relativity was offset with the help of increased overtime, a small amount of wage drift and, under external pressure again, the relaxation of wage ceilings in the higher-paid coalfields. Thus, in most coalfields, percentage increases in daywagemen's earnings remained on a par with increases in manufacturing, while the earnings relativity of pieceworkers was substantially eroded as aggregate wage control became increasingly effective in their case. In short, external wage movements set a ceiling on increases to daywagemen in coalmining and it was from this ceiling that internal wage differentials were contracted to produce a fall in the aggregate external relativity of the mining industry as a whole. While daywagemen continued to maintain their external relativity in the late 1960s and early 1970s, faceworkers fell even further behind other workers, outside and inside mining, as national agreements became more effective at the place of work. Thus the change in coal's market environment was felt most forcibly at the point of production.

So long as the industry's market position was weak and faceworkers' earnings varied between coalfields, it was difficult, despite growing discontent among facemen generally, to rally support to review the ever-increasing erosion of the faceworkers' wage relativity. Nor would one have anticipated much support from daywagemen if the miners' case had been based on a demand for an increased internal differential for faceworkers – a case for faceworkers had to be based on an increased relativity for all mineworkers, which was not difficult to make, so long as it was based on aggregate wage statistics which did not differentiate between faceworkers and other mineworkers. If successful, though, such a claim would inevitably hoist the external relativity of the lowest paid well above its previously highest post-war level. And this is what happened. Thus the previous chain of external comparisons was effectively broken. Faceworkers, upon whom increased production was most directly dependent, were now intent on recouping past losses and, given the internal wages policy of the NUM, taking the lower paid with them. And this policy of the NUM posed an increasingly severe threat to wage stability in the rest of the economy. Thus, mining turned from being a de-escalating to an escalating influence, which must become increasingly difficult to justify on grounds of special treatment, so long as the NUM sticks with its internal wages policy. Historically at least, a strong product market for coal has meant a widening of internal wage differential, in mining, and a weak product market has forced a reduction of differentials. Mining cannot regard itself as an island insofar as internal wages policy is concerned and it is certainly doubtful if a national incomes policy could withstand the repercussions of its doing so over a long period. So long as central incomes policy seems to bear unevenly on the public and private sectors, however, it will be difficult to guarantee the miners' support. For this, and other reasons, which will become apparent in subsequent chapters, it is not surprising that the industry should have considered ways of increasing the flexibility of its wages system.

PART THREE

PRODUCTIVITY, INDUSTRIAL RELATIONS AND LABOUR SUPPLY

9. THE NATIONAL POWER LOADING AGREEMENT

The replacement of piecerate payment forms in favour of fixed time rates in the late 1960s and early 1970s, and the reorganisation of the wage structure which followed that changeover, is undoubtedly one of the most interesting and hotly debated developments in coalmining wages policy. And in this context the National Power Loading Agreement of 1966 was of most significance.

Several arguments were advanced to justify the abolition of piecework in the negotiations which led up to the NPLA of June 1966. First, it was confidently expected that the changeover to day rates would reduce the frequency of strikes, and free management from the time-consuming task of negotiating price lists and settling disputes which arose from them. It was felt also that day wages would reduce the interminable wranglings which occurred when working techniques changed and encourage greater working flexibility. Second, there was a feeling that the daywage principle would encourage greater safety in the pits: there had been a temptation for pieceworkers to take unnecessary risks for the sake of meeting output targets and the NCB had been concerned at the rising accident rate. Third, it was argued that technology had reached a point where output was no longer determined by the physical efforts and skill of the men working the face and thus the piecework payment system was redundant.[1] From this, of course, it was only one step further to argue that the wages of power-loading men were really too high in relation to those of other miners. Fourth, the NCB was seemingly convinced that the wages of pieceworkers were still out of control[2] and that it would be difficult to improve the relative position of the lower-paid daywagemen so long as the industry retained a piecerate system.

We have already examined the impact of the NPLA on wage levels and movements, and suggested that wage trends *before* the abolition of piecework were not wholly compatible with the arguments used to justify the agreement. In this chapter we shall examine some other consequences of the NPLA and consider why the industry contemplated reintroducing some form of wage incentive scheme. As much of the support wage incentives have received in the 1970s revolves around the need to stimulate coal productivity, it is on this aspect of the debate that we focus most of our attention in this chapter. While the effect of abolishing piece rates on productivity is not easy to establish, our evidence does suggest that the change to fixed day wages impaired the rate of productivity growth. The impact of wages policy more generally on strikes and industrial relations is of sufficient interest and importance to warrant separate detailed analysis and we shall return to this question in the next chapter.

9.1 Coalface technology in the mid 1960s

The abolition of piece rates at the coalface, and the subsequent erosion of the faceworker's wage differential, was justified on the ground that technology had reduced the dependency of coal output on human endeavour and diluted the skills required of face teams. The substance of this argument deserves careful consideration.

The development of coalface technology has been outlined in earlier chapters (1 and 4). There we noted that, by the early 1960s, a stage had been reached where the completely automated coalface seemed a feasible proposition. While they were not completely automated, the ROLF installations at Newstead and Ormonde had done much to stimulate technological imaginations, and in 1965 Bevercotes hit the headlines as the pit of the future.

But the vision was unduly optimistic: automated power-loading machines which could cope with the diversity of conditions to be found at the coal face were still in their infancy, and much more research on automation was necessary. Moreover, although face-ripping and stable-hole machines had been developed, very few were operating by the mid 60s because of technical difficulties: stable holing and face ripping were still manual tasks in the typical pit. Again, powered supports were being introduced where face conditions were suitable for their use, but by 1966 these systems were in use on only 35% of power-loading faces: on the remainder props were still moved and set by hand. Thus, the only operations which were *generally* mechanised on a power-loading coal face in the mid 1960s were coal cutting and loading. It would hardly seem, therefore, that power-loading face output was independent of human effort in 1966, or that the men had no control over the rate of working.

The Board's technological argument was supported by the NUM. Indeed, as early as 1946,[3] the union referred to the industry's National Reference Tribunal a case for a national daywage agreement for power loading, on the grounds that it foresaw increasing difficulty in assessing the effort of the individual in the contribution to coal output on power-loading faces; in the NUM's opinion, even then, the end product would increasingly depend on machine efficiency rather than upon the individual effort of the face worker. On that occasion, however, the claim was turned down by the National Reference Tribunal[4] on the grounds that power loading was a 'development' of existing techniques rather than 'a revolutionary technique deserving of special attention'.

Again, at the NUM's 1963 Annual Conference[5] Will Paynter, the General Secretary, expressed the view that 'we have developed to a situation [on power-loaded faces] where it is the machine, and the sustained operation of the machine, that determines the level of production of most of our faces at the present time'.

The National Board for Prices and Incomes also supported the NCB's technological argument. In its 1966 Report on coal prices,[6] the NBPI remarked that piecework at the coal face had its origin in 'the fact that work at the face was in the past difficult to supervise and, since the work was performed by hand, an incentive was considered desirable in the interest of output. With growing mechanisation at the face this rationale will cease to hold good.' It went on to outline what it considered to be the defects of the existing piecework system and noted in passing that 'with mechanisation less physical effort is required for the work'.

The validity of the argument, that the skill and effort required on power-loaded faces was *less* than that needed on conventional faces, depended on which jobs one looked at. The task of guiding the machine certainly required less effort and skill than did steering the old coal cutter. Likewise, the job of cable handling (to help guide the machine) was a light and unskilled job – but this was fast disappearing with new techniques. Where powered supports were operating, the job of support setting was very light, if tiresome. The task of conveyor shifting, of course, was eliminated by this technique. Where supports were set manually, however, the task was unchanged from that under conventional mining, as was that of shifting the conveyor. Taking the above tasks as a whole then, there was something in the argument that the effort content had declined with power loading.

One might also conclude that the skill content had declined if one simply looked at the above tasks. But, because the Coal Board has insisted, since 1965, that all face men do a preparatory training course in basic face skills, one might reach the opposite conclusion if one applied the criterion to the *men* rather than the tasks. And one has to remember also that power-loading teams were handling very expensive equipment. On this account, one might conclude that the responsibilities of power-loading team members were greater than those of men on conventional coal faces.

So far, though, we have only covered one third of the jobs on a typical power-loading face in 1966. Other important tasks, now covered by the NPLA, were stable-holing and face ripping, which were unaffected by power loading. And it was the rippers and stable-hole men who largely determined the pace of the machine. The work in the stable holes and on the ripping lips was, and still is, extremely skilled and arduous. Face ripping, in particular, ranks as one of the most skilled jobs in the pit and for this reason, face rippers tended, on average, to work under better price lists (and earn higher wages) than did the other faceworkers: as early as 1963, for example, face rippers in Yorkshire received 6s 10d a shift more on their contracts than other faceworkers on power-loading faces. Nor were these men tied, like the other faceworkers, to a cyclical work pattern: they were very much their own bosses when it came to deciding how much work they should do. Thus, the more one examines the NCB's argument that machinery had reduced the effort, skill, and responsibility of jobs on power-loading faces by 1966, the more dubious it appears, and the more difficult it is to appreciate the support it received at that time.

9.2 Implementing the agreement

By any standards, the NPLA was a rushed job. The agreement was drafted and implemented in a few weeks. The agreed wage rates were, with the exception of Scotland and Durham, straightforward averages of EMS on power-loaded faces over the previous six months. But even then, the calculated NPLA wage rate was found to be an underestimate in at least one case. Management at division, area and pit level were not consulted on the details of the agreement. The first divisional and area management knew of the specific terms of the NPLA was when they received a copy of the agreement on 5 June 1966 with instructions to implement it on new faces the next day. And the guiding notes on interpretation from the NCB in London were not received in the Divisions until 7 June – one day *after* many of them had implemented the agreement.

This writer can confirm that on the night of 5 June 1966 one could have visited the homes of divisional wage control officers to find them, in the absence of national directives, drawing up their own hurried notes to guide local managers on the agreement. In the absence of central guidance, it was hardly surprising that interpretations of specific points of the agreement were often contradictory, or that further anomalies were discovered when the national 'Notes for Guidance' were received.

When the NCB did eventually call a conference of divisional wages and control officers to compare guidelines, it was some two weeks after the agreement was in operation. At that meeting many of the national directives were shown to be impracticable and had to be altered, as did many of the local interpretations. But even then many of the problems were not ironed out. Thus, because of bad planning, the NPLA got off to a bad start.

The way in which the NPLA was introduced is described here, not so much as a criticism of the NCB – although it is undoubtedly difficult to reconcile the above description with the NCB report[7] that the agreement was introduced smoothly – but rather because it may be relevant to the reaction to the agreement in the coalfields.

9.3 The effort bargain

We traced the impact of the NPLA on pay levels and differentials in earlier chapters. At this juncture, though, it is necessary to reiterate some of its short-run implications for pay before moving on to consider the effects of the agreement on productivity.

The impact of the NPLA on pay in the short term varied between coalfields and between the pits within them. In Scotland and Durham the NPLA wage rates were either equal to or slightly above the existing daywage rates payable under district agreements, and for this reason all power-loading team members in these coalfields were placed on the NPLA as from 6 June. A more gradual approach occurred in other coalfields because of the wide disparities in shift earnings under piecework agreements. In coalfields other than Scotland and Durham, therefore, the NPLA was to apply only on new coalfaces, opened on or after 6 June. Thus power-loading teams would continue to be governed by district wage agreements until they were transferred to a new coalface. In the meantime, however, the NPLA rate became the effective minimum below which the district rate could not fall. On being transferred to a new coalface any team member who had previously earned more than the NPLA rate suffered a wage cut. The actual implications of the NPLA on the district wage level thus depended on the speed at which new coalfaces were opened and the wage level prevailing on coalfaces being closed down. The proportion of men receiving a wage cut when transferred to new coalfaces would depend on the skewness of the existing wage distribution. The writer has not been able to obtain data on the distribution of EMS for all coalfields. The general impression obtained, though, is that the NPLA wage rates – determined as the mean EMS on power-loading faces during the six months before June 1966 – implied a higher proportion of power-loading team members suffering wage cuts on transference to new faces in the East Midlands and Yorkshire coalfields than elsewhere; and particularly so in the longer term as, to achieve parity, wage rates in these coalfields were to rise less fast than in the lower-paid coalfields.

Nor was the impact of the NPLA on wage rates of different jobs covered by the agreement by any means uniform. Because of their relatively high shift earnings, face rippers who were now included in the power-loading team tended to suffer most under the NPLA. In Yorkshire, for example, the distribution of shift earnings between face rippers and men in the power-loading teams just before the NPLA was such that approximately 60% of power-loading operators received an increase in their shift rate, compared with only 40% of rippers.

Craftsmen were the group to benefit most from the NPLA. Thus the shift rates of fitters and electricians who were attached to a power-loading team operating under the NPLA rose overnight from the standard grade rate of 54/- a shift to the appropriate power-loading rate, which varied from 75/- to 88/11 a shift. But animosity was also created within the craft ranks. First, not all craftsmen were to get the power-loading rate. Second, because NPLA rates were not uniform between the wage districts, some craftsmen who were included in face teams received larger increases than others. Generally, though, craftsmen were keen to see new coalfaces opened, whereas in the higher-paying pits the pieceworkers, and the rippers in particular, were not. This also created some friction between the two groups.

Let us now turn to the effects of the agreement on productivity. Here we shall confine our attention to the first year or so of the agreement's operation, as the considerable changes which have since occurred in the industry make it difficult to come to meaningful conclusions thereafter.

The NPLA applied only to new power-loading coalfaces starting on, or after, 6 June 1966. As no two coalfaces operate under the same geological conditions and new faces probably incorporated techniques not found on old faces, it is very difficult to trace the impact of the agreement on productivity with much accuracy. Furthermore, as productivity is measured by dividing output by the number of manshifts worked, the effect of including craftsmen for the first time as full-time members of the power-loading team on production shifts would *ceteris paribus* reduce face OMS even if actual face output remained constant.

This latter difficulty makes conclusions about the effect of the NPLA on productivity based on crude statistics of *coalface* OMS extremely dubious and, in the short run at least, the question is best analysed on the basis of trends in overall OMS. In Figure 9.1,

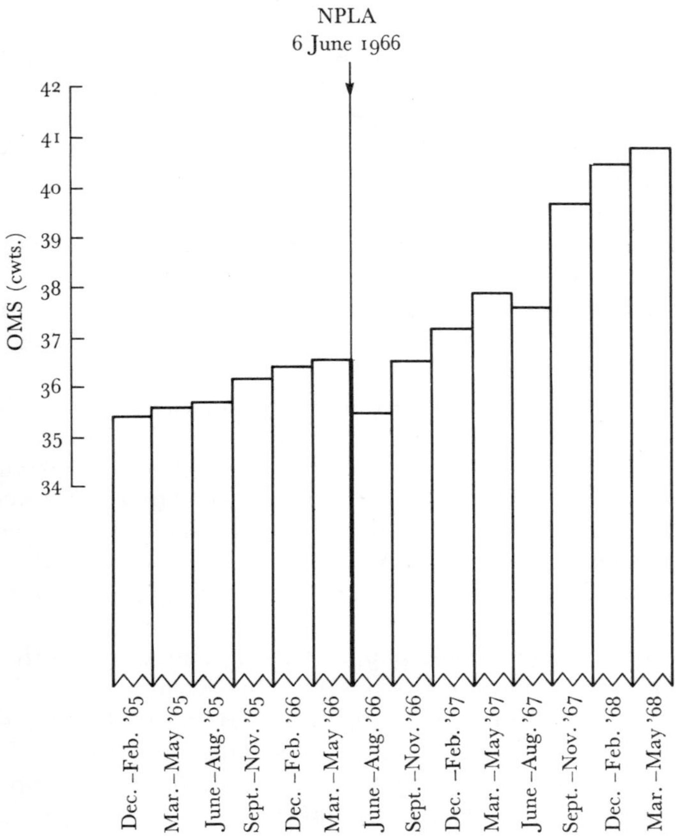

Figure 9.1 Quarterly output per manshift (deseasonalised) of all mineworkers
Source: CSO, *Monthly Digest of Statistics*.

therefore, we have traced the course of overall OMS, adjusted by conventional statistical techniques to eliminate the strong seasonal component in productivity, for quarters before and after the introduction of the NPLA.

There was clearly a tendency for OMS to increase less fast in the eight months prior to the agreement than hitherto. This might be explained by a slowing down in the percentage increase in the amount of coal cut by power loading during that period. However, the most notable feature of the diagram is the sharp fall in OMS in the first quarter following the implementation of the agreement. And, while OMS had apparently returned to its immediate pre-NPLA level by the second post-NPLA quarter and surpassed it by the third, this was partly due to the considerable improvements in transport and haulage arrangements behind the coalface. Furthermore, the increase in overall OMS by the end of 1967 was partly due to the more rapid closure of uneconomic and less productive pits. This has been estimated (by the NCB)[8] as improving overall OMS by about 1.5% per annum (0.7 cwts of coal per shift) between 1966 and 1968.

The picture is even less favourable when one remembers the underlying trend of coalface technology since the agreement was implemented. Thus, between June 1966 and June 1968, the proportion of faces with self-advancing powered supports rose from 35% to 75% and the number of stable hole and ripping machines also increased. Given that the NCB estimated OMS on power-supported faces as being some 30% higher than on non-power-supported faces, the true picture by 1967 is probably much less favourable to the NPLA than the OMS figure suggests. It is this writer's guess therefore that, allowing for the effect of technical change and pit closures, productivity, having fallen quite substantially in the months immediately following the NPLA, did not return to its March–May 1966 level until the third, or possibly the fourth quarter following the agreement.

While it is no proof of this supposition, a comparison of the NCB's own projected estimates of OMS with actual results, set out in Table 9.1, tends to confirm the disappointing trend of OMS after the NPLA. And one should note that the discrepancy between actual and budgeted OMS was three times as great in 1966–67 as in subsequent years, when the NCB was able to make a more realistic estimation of budget OMS in the light of the new circumstances. In fact the level of OMS budgeted for in 1966–67 was only just realised in the following year.

Having traced what appears to have been a deleterious effect of the NPLA on productivity in the industry as a whole in the short run, let us now turn to the position in the various coalfields. Again, due to the difficulty of obtaining a reliable index of face productivity, this

Table 9.1. *Budgeted and actual productivity comparisons*

Year (June–May)	Budgeted OMS	Actual OMS	Difference
1966–67	38.3	36.6	−1.7
1967–68	39.6	39.0	−0.6
1968–69	42.9	42.5	−0.4

Source: National Coal Board

analysis is confined to an overall OMS statistic. For reasons already outlined, it should also be remembered that these figures tend to give a more favourable impression of productivity movements than would those of face OMS.

In columns (b), (c) and (d) of Table 9.2 the deseasonalised quarterly statistics for the three quarters following the agreement are expressed as a percentage of the OMS recorded in the quarter immediately preceding the NPLA (column (a)). With the exception of four wage districts (Kent, South Derbyshire, North Wales, Lancashire and Cheshire) where OMS had been fairly static, OMS had been rising steadily in the months immediately before the agreement. With the exception of five wage districts (South Wales, South Derby, North Wales, Durham, and Scotland), however, OMS declined during the first quarter following the introduction of the NPLA. In some areas the decline in OMS was slight (less than 2%),

Table 9.2. *Quarterly trends in overall OMS in the wage districts*

	OMS before NPLA			OMS after NPLA			
	(a)	(b)	(c)	(d)	(e)	(f)	(g)
Wage district	March –May 1966	July –Sept. 1966	Oct. –Dec. 1966	Jan. –March 1967	b ÷ a	c ÷ a	d ÷ a
Kent	29.6	29.3	30.1	30.9	99.0	101.7	104.4
South Wales	25.8	26.1	26.3	26.8	101.2	101.9	103.9
Warwick	39.9	38.1	38.8	41.0	95.5	97.2	102.8
Shropshire	33.9	33.0	33.7	31.9	97.3	99.4	94.1
South Staffs.	26.2	20.5	19.6	20.0	78.2	74.8	76.3
Cannock Chase	44.2	41.5	42.8	41.4	93.9	96.8	93.7
North Staffs.	40.0	39.7	40.9	41.1	99.3	102.5	102.8
Leicester	59.5	59.0	57.7	60.1	99.2	97.0	101.0
South Derby	52.9	53.6	51.5	50.9	101.3	97.4	96.2
Notts.	53.7	51.6	50.6	51.6	97.0	95.1	97.0
North Derby	45.6	44.0	47.0	47.0	96.5	103.1	103.1
North Wales	23.8	24.0	25.6	25.0	100.8	107.6	105.0
Lancashire & Cheshire	32.1	31.4	30.6	30.9	97.8	95.3	96.3
Yorkshire	38.4	37.1	35.1	39.0	96.6	91.4	101.5
Durham	28.5	29.4	29.6	32.3	103.2	103.8	112.9
Northumberland	36.2	35.5	35.2	33.5	98.1	97.2	92.5
Scotland	31.7	31.9	31.8	33.2	100.6	100.3	104.7

Source: National Coal Board

but in others (Warwick, Shropshire, South Staffs., Cannock Chase, Nottingham, Lancashire, Yorkshire, and Northumberland) it fell by 2% or more. The largest decline was in South Staffs. (21.8%). By the third quarter, following the introduction of the NPLA, OMS in eleven of the wage districts was equal to, or higher than, that recorded in the quarter immediately prior to the introduction of the agreement. However, in seven districts (Shropshire, South Staffs., Cannock Chase, South Derby, Nottinghamshire, Lancashire and Cheshire, and Northumberland) OMS was still below that level. OMS improved fastest after the agreement in Durham and Scotland. Again, one must interpret these statistics with some caution. Productivity growth had already shown signs of being slower in the higher-productivity areas than in the low-productivity areas, as the scope for additional use of power-loading machines became increasingly confined to coalfields in the latter group. But this factor does not explain why OMS actually fell in some coalfields in the months immediately following the NPLA. And, whereas technological trends may provide part of the answer, it seems more than a mere coincidence that two of the highest productivity improvements following the NPLA were in the two coalfields – Durham and Scotland – which previously had district power-loading agreements most akin to the NPLA, and where most faceworkers received some increase in wages following the latter's implementation.

It is also interesting to note in Table 9.3 that, among the coalfields previously using piece rates at the coalface, productivity seems to have been most adversely affected in the West Midlands and Northumberland coalfields which, in March, 1967, had the highest percentage of faces operating under the NPLA, and that among the

Table 9.3. *The percentages of coalfaces in each Division on the NPLA in March 1967*

NCB Division	% of faces on NPLA
Scotland	100
Durham	100
Northumberland	48
Yorkshire	34
N. Western	27
E. Midlands	32
W. Midlands	53
Wales	11
Kent	47

Source: National Coal Board Information Paper 34

coalfields where productivity was least affected we should find South Wales, which had the lowest percentage of faces operating under the NPLA.

If, as seems realistic, one can assume that the proportion of face team members receiving a wage cut was positively related to the number of faces conditioned to the NPLA, this suggests that as face teams moved to new faces they adjusted their effort input in response to what for many would have been reduced wage rates under the NPLA.

As conditions vary enormously between coalfaces, productivity comparisons between faces operating under the NPLA and those on district wage agreements must be treated with extreme caution. Nevertheless, Table 9.4, based on data supplied by the NCB, illustrates some aspects of performance on power-loading faces working under the NPLA and those not working under the agreement in March 1967.

For the purpose of this analysis, NCB Areas have been divided into high, medium and low productivity groups: to have conducted the analysis on a national basis would have given a misleading impression because of the high proportion of faces on the Agreement in Durham and Scotland, where productivity has traditionally been well below the national average.

The table shows that OMS in each range was appreciably lower on NPLA faces than on other faces which were operating under local or district agreements, although daily output was highest on NPLA faces, as a result of a higher rate of machine shifting. Also, while OMS was lower, the number of square yards cut per machine shift on NPLA faces was generally much higher than on other faces. And, even in the high OMS group, where the square yardage comparison is slightly in favour of the other faces, the difference bears little relation to the unfavourable productivity comparison. The reason for the lower OMS on NPLA faces is to be found in the last column of the table. This shows that in all three productivity groups more manshifts were worked per machine shift on NPLA faces than on the older faces, particularly in the medium OMS areas. In short, the better machine utilisation achieved on NPLA faces, and subsequently the higher output, were obtained through a disproportionate increase in the number of men employed on those faces. Thus, one reason for the fall-off in OMS after the NPLA was introduced would appear to be that manpower complements on NPLA faces tended to be higher than on existing faces. This may be attributable, in part, to the inclusion of craftsmen in face teams, but it was certainly not the only factor serving to increase manshifts per machine shift; these trends

Table 9.4. *Some aspects of face performance in March 1967*

	No. of machines	Manshifts per day	Pithead OMS (cwt)	Pithead output per day (tons)	No. of machine shifts per day	Sq. yds per machine shift	Manshifts per machine shift
High OMS areas[a]							
NPLA faces	180	11,284	200.6	628	2.22	186.4	28.3
Other faces	327	18,430	219.5	619	2.15	189.0	26.3
Medium OMS areas[b]							
NPLA faces	98	7,891	157.1	633	2.50	170.3	32.2
Other faces	160	11,238	179.5	630	2.49	157.3	28.2
Low OMS areas[c]							
NPLA faces	109	8,422	130.1	502	1.86	148.1	41.4
Other faces	175	12,086	139.1	480	1.80	143.9	38.3

[a] North Nottinghamshire, North Derbyshire, South Nottinghamshire, South Midlands, North Yorkshire, Staffordshire.
[b] Doncaster, South Yorkshire, Barnsley, Northumberland.
[c] North Western, Kent, East and West Wales.
Source: National Coal Board

must also be set against the fact that face technology was veering towards labour displacement at the face ends and support setting, especially so in the high and medium OMS areas. At this point it will be convenient to examine some of the problems encountered by local management in operating the NPLA.

We emphasised earlier the dependence of face advance on the stable-hole men and the rippers. One of the main complaints after the introduction of the NPLA was that the ripping and stable-hole teams were failing to keep up with the advancing faces. Many of these men had been earning much more than what became the NPLA shift rates and were not unnaturally reluctant to maintain their effort input at a reduced price. This was particularly true of the ripping teams. But previously the rippers' higher wage had been obtained by rather devious means. They would bargain on the price to be paid a given number of men in the team. It was in the rippers' interest, of course, to get as many men in the team as possible when the price was struck. Once the price had been fixed, however, the teams would often shed one or two of their number so that the price paid would be shared between fewer men and thus raise the average wage of the remaining members. The NPLA provided no incentive for the rippers to bargain in this way: the men now wanted to get the maximum number of men in the team to lighten the workload of individuals, but there was no incentive for them to subsequently lighten the team. The nature of ripping, and the physical conditions under which it is performed, give a very definite upper limit to the number of people one can put in a ripping team – usually five or six at the most. When a new face went on to the NPLA, the size of the ripping team would be determined, in the running-in period, by reference to manning on faces elsewhere. But if the rippings fell behind, there was a tendency for management to raise the number of men in the team to its limit to achieve maximum machine utilisation.

Nor was ripping the only task which lent itself to additional manning. Only one man is required on any power-loading installation to guide the machine and, on most, only one more to hold the guiding rope. There is no scope for additional manpower on these tasks. But when the NPLA was introduced, power loading was at a stage of development which left plenty of scope for bargaining over the number of support setters required – and even on powered-supporting systems there was some flexibility in this respect. In several pits which the writer visited during the early stages of the NPLA, a tendency towards extra manning on supports setting was reported by pit managers: whereas it was previously recognised that one man would look after about 13 yards of support on manual support systems, this had decreased to some 11 yards on many NPLA faces.

But why, one might ask, were these problems not overcome by the method study provision of the agreement? The answer is twofold. First, the agreement was based on the daywage principle: there was no mention of task norms. Method study was applied to the machine to determine manning with a view to maximum machine utilisation. That was very different from deciding what the machine and the men were capable of doing and fixing *task norms* accordingly. We have already noted the opposition within the NUM to method study of the men. And it is interesting to note that Will Paynter, sensing the opposition which any mention of method study might provoke from delegates, stressed the inoffensiveness of the NPLA's method study clause at the Special Conference called in 1966 to discuss the agreement. Second though, and equally important, the method study provision was not really applied systematically in many areas; more often than not, manning was decided by local custom and practice. Not surprisingly, perhaps, manning was most rigorously controlled in those areas which could least afford to carry a surplus of men.

So far we have examined some of the problems of operating the NPLA on new faces. But the agreement also affected old faces. Although power-loading teams on coalfaces which were already operating on 6 June 1966 were not brought under the NPLA, it was agreed that the NPLA shift rates would become minima below which the shift earnings, derived from district wage agreements, would not fall. Thus overnight the incentive element in many of the district agreements was reduced to negligible proportions. Output could fall below agreed norms, and so long as management could not identify an organised go slow (which was dependent upon there being an issue in dispute), they had to pay the NPLA rate, rather than the Grade I underground rate. In addition, men knew that as soon as the coalface on which they were working finished, they would be put onto the NPLA shift rates. In those pits where the average EMS was below the power-loading rate on district agreements there was no real incentive to finish the face to obtain the higher rate. On those faces where the district EMS was higher, on average, than the NPLA rate there would be a natural tendency to work more slowly to prolong the life of the coalface, but not so slowly as to reduce EMS to the NPLA rate.

In addition to the above problems, many managements reported difficulties in ensuring an adequate flow of supplies to the face. When the power-loading teams were operating under wage incentives it was in their interest to ensure that supplies reached the face as they were required and to ensure that technical difficulties were corrected as soon as possible. If the supervisor was not available when the problem occurred, the men would either take the necessary action themselves or make every effort to find him. With a daywage payment system, this incentive disappeared. Thus supervisors found themselves having to keep a very close watch on the coalface to prevent delays in supplies, which seem to have become a major problem. Thus the NPLA considerably increased the pressures on underground supervisors, who, as one pit manager put it, 'had to start doing their job'.

While the percentage of supervisors to face personnel had been increasing steadily before the introduction of the NPLA, the NCB reacted to the emphasis which the agreement placed upon efficient supervision by further increasing the proportion of supervisors at the coalface. In 1966, for example, there was one supervisor to just over five faceworkers. By 1968, there was one supervisor to just over four faceworkers. The ratio has fallen still further since. Of special importance in this trend was the creation of a new grade of face superintendent on power-loading faces, to help with the difficulties outlined above.

It is exceptionally difficult to evaluate the effect of the NPLA on productivity in the longer term. In the accounting years 1968–69, OMS increased by 9% over the previous year, although it was still below the budget forecast. This might imply that some of the difficulties outlined above had been overcome. By the year 1974–75, however, OMS had only increased by a further 6% on its 1968–69 level. The upsurge in productivity in 1968–69 may certainly be regarded as exceptional. One reason for the sudden improvement was that management neglected development for the sake of concentrating manpower on coalface production. This led to problems later and helps to explain the poor productivity performance in subsequent years. The slow growth of OMS after 1969 was also the consequence of attempting to reduce the number of faces in each pit to cut costs. This resulted in a shortage of spare faces to which labour could be deployed when machines on operational faces broke down. There had also been a tendency, up to 1973, to increase the number of machine shifts worked per day. As a consequence, machine maintenance was neglected and mechanical breakdowns increased. That trend has since been reversed. In addition, the industry was severely disrupted in 1972 and 1974 by national strikes which were themselves not unassociated with the decline in the miners' relative wage advantage as a consequence of the wage structure revisions of 1966 and 1971.

Important as all these factors undoubtedly are in explaining the industry's poor productivity record in the 1970s, the evidence also points to a substantial productivity reserve. In the late 1960s and early 1970s, the industry was not achieving more than 60% of method-study output. This, in turn, encouraged local management to fall back on its own initiative to raise output and earnings opportunities. One of the most frequently adopted tactics was to increase the proportion of coal cutting done in overtime hours. Thus, whereas overtime shifts were only 4.2% of total shifts worked at the coalface in the period 1963–65, the proportion had increased to 8.6% in 1969.[9] Despite attempts to reduce overtime, it was still running at 7% in the early 1970s. As the NCB applied pressure to reduce overtime, however, pit managements evolved their own incentive schemes. One delegate at the NUM 1970 Annual Conference[10] commented on the situation as follows:

'The most ridiculous things are happening at the pits. In fact Dr Johnson and Captain Cook would probably be jealous of the attitude of the Board at the present time in offering bribes known as incentives to the lads that would probably be worse than Captain Cook's offering of mirrors and beads to the natives when they landed on foreign soil. In our Area [Midlands] we have the situation of barrels of beer being on racks when they ended a shift for the lads who have reached a certain productivity figure, chicken at Christmas, breakfast, dinner and supper tickets that would require an accountant to count them up at any pit as a bribe to stop overtime... These are the things that are actually happening.'

There can be little doubt that many of the problems associated with the changeover from piece rates to time rates were exacerbated by the declining internal and external wage relativity of the groups affected. One consequence of this was the increased dependency of the NCB on outside contractors, who paid much higher rates of pay than those specified for NCB employees, for tunnelling and roadway ripping. This caused considerable resentment inside the NUM. And the fact that many of those firms would employ ex-NCB miners, and achieve much higher rates of productivity than the NCB's own miners was frequently cited as a reason for reintroducing wage incentives on roadway development.

9.4 Accidents, disputes and local bargaining

Before we consider early attempts to reintroduce wage incentives into the industry, there are one or two other consequences of abolishing piece rates which deserve attention.

The abolition of piecework, and with it some of the pressures on men to take unnecessary risks, may have been instrumental in reducing the incidence of underground mining accidents. Our data in Table 9.5 show that, having increased from 188.2 to 243.5 per

Table 9.5. *Underground accidents at NCB mines* (rates per 100,000 manshifts)

	1962	1965–66	1967–68	1970–71
Falls of ground – face	39.9	46.1	39.8	25.4
Falls of ground – roads	4.3	5.1	5.2	5.0
Haulage and transport	23.0	25.3	20.7	13.3
Machinery	3.5	5.6	6.8	8.0
By use of tools and appliances	17.0	21.8	20.8	14.1
Stumbling, falling and slipping	31.4	45.9	44.3	32.3
Other causes [a]	69.2	93.7	84.9	61.0
Total	188.2	243.5	222.5	159.0

[a]The major causes in this general group are through falling objects and handling supplies.
Source: NCB, *Reports and Accounts*, Volume II: accounts and statistical tables (1962 – 1970–71)

100,000 manshifts between 1962 and the accounting year 1965–66, the accident rate fell quite substantially, to 222.5 in 1966–67 and 159.0 in 1970–71. As the NCB does not distinguish, except in the case of accidents caused by falls of ground, between accidents which occurred at the coalface and elsewhere underground, it is difficult to say how much of the improvement comes from the coalface. If the pressure had been taken off coalface workers to work negligently, we might, for reasons already explained, expect part of the effect of a slower, but safer, speed of working at the face to be transmitted to jobs elsewhere underground, such as haulage and transport. In the case of accidents caused by falls of ground at the face, the general trend may be thought favourable to the agreement, in that, after increasing quite substantially from 39.9 to 46.1 per 100,000 manshifts, accidents from this cause at the face subsequently fell continuously to 25.4 per 100,000 manshifts in 1970–71. In part, this trend undoubtedly reflects the widespread adoption of powered-support systems in the late 1960s, which offered more physical protection from falling material than did the more conventional systems. Important as the technological improvements were, we might have anticipated some reduction in the accident rate from this cause before 1966 as well, but of this there is no evidence. Although we might have expected some improvement in the accident rate after 1966 anyway, one is tempted to believe that the abolition of piecework played some role in reducing the accident rate.

It was expected that the abolition of piecework would reduce the frequency of strikes and encourage greater working flexibility. We shall analyse the impact of wages policy on industrial relations, and strikes in particular, in the next chapter. Suffice it to say here that the NPLA (as well as the revised elsewhere underground pieceworkers wage structure) seems to have produced a significant shift in emphasis in local bargaining. Our data in Table 9.6, for example, show that the total number of work stoppages in coalmining declined from 1,401 in 1963 to 473 in 1972. Although the number of stoppages had fallen quite dramatically over the two years prior to the introduction of the NPLA in 1966, the proportion of stoppages over wage issues actually increased from 41.5 per cent in 1963 to 45.1 per cent in 1965. By 1972, however, the proportion of stoppages over wage issues had fallen to only 13 per cent. Total work stoppages declined by 570 between 1965 and 1972. Of this reduction, 82 per cent was attributable to the decline in stoppages over wage issues.

It is particularly interesting, in the light of our previous discussion, that, despite the sharp reduction in the number of miners employed, the number of work stoppages caused by refusals to accept alternative work increased between 1967 and 1972, and almost reached the level

Table 9.6. *The cause of work stoppages 1963–72*

Reported cause of stoppage (NCB definitions)	Number of stoppages			
	1963 No. (%)	1965 No. (%)	1967 No. (%)	1972 No. (%)
Allowances and bonuses	87(6.2)	53(5.1)	37(6.5)	19(4.0)
Wage and price lists (other than above)	494(35.3)	417(40.0)	138(24.2)	41(8.7)
Five-Day-Week Agreement	5(0.4)	1(0.1)	11(2.0)	7(1.4)
Refusal of alternative work	97(6.9)	86(8.1)	66(11.6)	92(19.5)
Refusal to perform work left over from previous shift	17(1.2)	22(2.1)	12(2.1)	8(1.4)
Refusal to repair after mechanical breakdown	4(0.3)	1(0.1)	2(0.4)	2(0.4)
Methods of working or organisation	301(21.5)	263(25.2)	150(26.4)	107(22.6)
Bad working conditions	147(10.5)	82(7.9)	65(11.4)	50(10.5)
Personnel or grading	46(3.3)	23(2.2)	19(3.3)	4(0.8)
Objections to officials	41(2.9)	10(1.0)	21(3.7)	25(5.2)
Sympathy with men dismissed or reprimanded	53(3.8)	14(1.3)	28(4.9)	34(7.1)
Retaliatory	7(0.5)	11(1.0)	9(1.7)	25(5.2)
Closures, redundancy, transfers	2(0.1)	4(0.4)	1(0.2)	1 –
Miscellaneous	100(7.1)	56(5.4)	10(1.8)	58(12.3)
Total number of stoppages	1,401	1,043	569	473

Source: National Coal Board

of 1963. Thus, whereas work stoppages over wages declined from 1.00 to 0.2 per thousand people employed between 1965 and 1972, work stoppages over refusals to accept alternative work increased from 0.19 to 0.34 per thousand people employed, over the same period. The latter trend can be accounted for by the refusal of men to perform the more arduous face tasks at a wage rate which, being uniform, did not discriminate between the various types of work covered by the NPLA. If these statistics are anything to go by, it would seem that the NPLA met with less success than anticipated in encouraging greater flexibility of working at the coalface.

While the NPLA and the subsequent revision of the elsewhere underground pieceworkers' wage structure between 1969 and 1971 seems to have resulted in a shift of emphasis in the cause of local disputations, those agreements considerably reduced the workload of lodge officials and reduced their significance in the eyes of their members.

Whereas lodge secretaries always tended to complain about poor attendance at branch meetings, their complaints on this account certainly increased after the introduction of the NPLA. Indeed, one lodge secretary went so far as to tell the union's Annual Conference[11] that he and others like him might just as well be replaced by NUM social insurance officers, as social insurance problems were now the main item of local business. It would seem, then, that many lodge secretaries felt their role and status in the community had been severely eroded by the abolition of piecework in the pits. Against this, however, the feeling disappeared that the NUM lodges existed mainly for the benefit of the contract workers. As Lawrence Daly put it:

'The main criticism that I have had in all my experience in the union from our lower-paid daywagemen and from the men on the surface is that this is not their union, it is the piece-rate workers' union, because the local branch officials, where a piecerate system prevails, spent 90 per cent of their time servicing the men who are on local contracts.'

A large proportion of the National Executive Committee were, of course, aware that to reintroduce wage incentives might weaken the NUM's central authority. This proved a major stumbling block in subsequent negotiations on the subject.

9.5 Wage incentives: the debate 1972–77

The suggestion that the industry should consider some form of wage incentive scheme first received public support from the Wilberforce

Enquiry[12] – established in 1972 to settle the miners' strike that year. It was subsequently endorsed by the Pay Board and the Coal Mining Industry Examination Report[13] of 1974. The NCB and NUM, however, were divided on the type of scheme which would best serve the industry.

Both sides agreed that they did not want a return to a straightforward piecerate system. The main argument revolved around how wage incentives should be distributed and the basis on which incentives should operate. The NCB favoured a pit-based scheme biased towards direct production workers, while the NUM, fearful of the divisive effect on their own organisation, preferred a national scheme which would spread the rewards more equally among the labour force.

In September 1974, the NCB proposed a scheme designed to achieve an output target of 120 million tons of coal. It would have operated on a national basis but been linked with local incentives. It proposed that standard tasks and manning be established by method study for each face and development drivage. The tasks were to be expressed in square yards of coal cut at the face and lineal yards of advance in development drivages. The basic task would have been set at 75% of standard and paid at the national shift rate. An extra £2.50 per shift would have been payable if the standard was reached and, thereafter, the bonus was to be 10p per shift for each additional 1% by which actual performance exceeded the standard. That scheme would have applied to production workers at the face and development headings. Other faceworkers, and grade A workers, were to receive a bonus calculated as the average of bonus payments on all installations (faces and drivages) at the pit. Other miners were to receive a bonus equal to 55% of the weighted average of all incentive payments made on installations in a particular *area*.

The NUM suggested several amendments to the scheme. First, it proposed that the general bonus be calculated on a *national* basis, and fixed at 65% of the national average of incentive payments. Second, that the bonus for face teams should be calculated on a pit, rather than on a face, basis. Third, that there should be a separate bonus for development work, which would be paid on a team basis. The idea was to raise the wages of development teams to a level comparable with those earned by outside contractors and, by making development work more attractive to existing employees and increasing the cost of not using them, discourage the employment of outside contractors.

Despite the NCB's acceptance of the NUM amendments, the

union's executive, under pressure from the leftwing areas, recommended its members to reject the scheme. This they did by a vote of 61.5 per cent against and 38.5 per cent in favour. As a result a national production bonus scheme was introduced in early 1975 instead of the pit scheme. A bonus was to be paid under the national scheme if overall output of the industry exceeded an established target. The bonus was to be the same for all miners. During the first three months of the bonus scheme output did exceed the target and the NCB paid a bonus of £2.90 per week to each man in the industry. Thereafter, output consistently fell below target and the bonus was withdrawn. Although the scheme continued to exist, negotiations were reopened on a more decentralised pit scheme, which would have related the bonus of production workers to individual coalface task standards, but paid a uniform bonus to all other workers based on the average bonus of each pit.

Although, in theory, it should be possible to establish consistent task standards between coalfaces, the practical problems are clearly enormous, and the NUM's executive has understandably been concerned in case some areas benefit more than others. As task norms are almost bound to be set with some regard to existing custom and practice, there must be a danger that the pits and areas with the greatest productivity reserves will benefit most. As it now seems fairly certain that some form of wage incentive scheme will be introduced in the near future, in an attempt to boost coal output and productivity, a wage structure is likely to emerge with rather more variation between coalfields and between direct and indirect production workers in the industry. This could well weaken the NUM's central authority over its constituent parts. The most important background factor to the discussion of incentive schemes was the need for more coal. It might be more than pure coincidence, then, that the decision to abolish piecework was originally taken in the opposite circumstances.

9.6 Summary and conclusions

Many of the arguments used in 1966 to justify an abolition of wage incentives at the coalface appear rather dubious. The agreement resulted in reduced productivity as face teams adjusted their effort input to the often reduced wage rates. The switch from wage to effort bargaining was also reflected in an increased incidence of disputes over refusals to accept alternative work and higher overtime levels at the coalface. These problems necessitated more coalface supervision and encouraged pit managers and lodge officials – many of whom resented the erosion of their local bargaining authority – to experiment with alternative 'incentive schemes'.

The swing of academic opinion in the mid 1960s against piecerate systems played an important part in the sudden change of opinion inside the NCB in favour of daywage payment systems at the coalface. But the idea also gained acceptance in the coal industry because a national daywage agreement would give the NCB and NUM greater authority over wage determination at a time when the market for coal was diminishing. Just as the NPLA was compatible with the planned rundown of coal production, so a return to local wage incentive schemes has been encouraged by the strengthening of coal's market position in the 1970s. Certainly attitudes to wage systems in coalmining have not been immune from product market influences.

10. WAGES POLICY AND INDUSTRIAL RELATIONS

We touched very briefly on some effects of wage reform on coalmining industrial relations in the previous chapter. But this subject clearly demands more rigorous analysis. Of particular interest is the hypothesis of some academics[1] that changing wage-payment practices at the coalface before the NPLA influenced the sharp decline in recorded work stoppages in mining around the late 1950s. In this chapter we examine this hypothesis in some detail, with the help of hitherto unpublished data, and comment on more recent strike trends in the industry. Our analysis suggests that, while wage reform certainly had some effect on the industry's strike record, it constitutes only a partial explanation of the sharp decline in strike activity in the late 1950s and early 1960s. And whereas the later revisions of the piecework structure reduced the frequency of work stoppages over wage questions, this must be seen in the context of the changing character of strike action, which is associated with the impact of wage reforms on wage trends and the more unified approach to wage problems which they facilitated.

This chapter is divided into four parts. First, there is a brief examination of the trend in recorded work stoppages in coalmining and 'all other industries' combined since 1920. Second, the plausibility of the 'change in wage payments systems' hypothesis is examined. Third, an alternative explanation of the declining frequency of coalmining strikes around the mid 1950s is offered. We conclude with a brief analysis of strikes in the early 1970s.

10.1 The trend of work stoppages

The number of recorded work stoppages in coalmining and 'all other industries' combined between 1920 and 1974 is shown in Figure 10.1, as is the percentage share of coal stoppages in the total recorded for all industries combined. The importance of the graph is twofold. First, it shows quite clearly that, although the share of coalmining stoppages in the total recorded increased from 12% in 1920 to 78% in 1957, the frequency of stoppages in coalmining *varied* rather similarly to the frequency of stoppages in all other industries over that period. After 1957, however, the stoppages recorded in coalmining *fell*

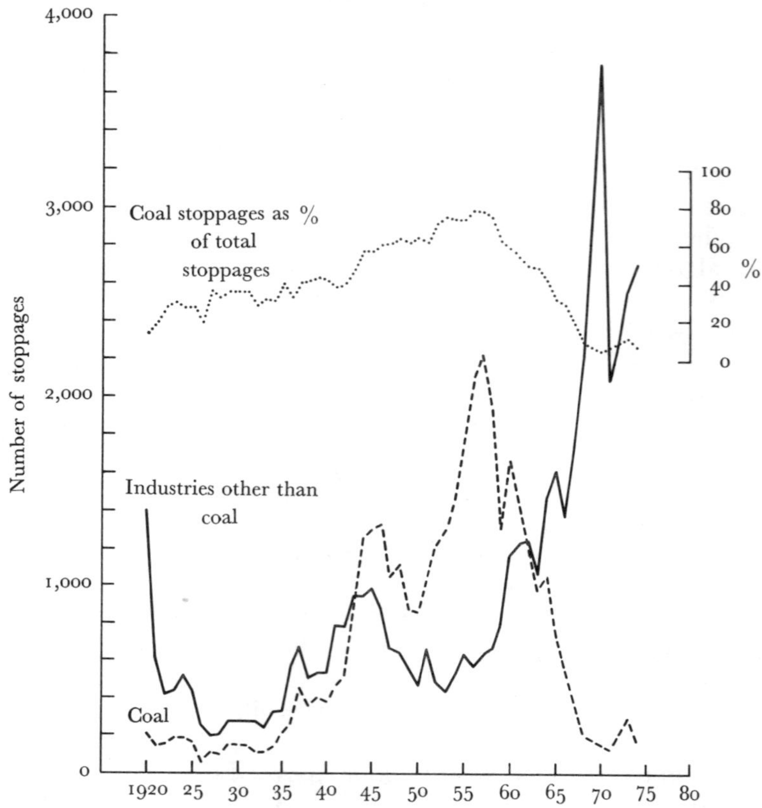

Figure 10.1 The number of reported work stoppages in coalmining and 'other industries' 1920–74
Source: Department of Employment, *Gazette*.

quite dramatically, (from 2,228 in 1957 to a post-war low of 221 in 1970), while the number in all other industries combined rose from 860 to 3,888 over the same period: by 1970, after which the number of mining stoppages rose slightly, coal accounted for only 4% of the total work stoppages recorded by the Department of Employment. Thus, whereas variations in the frequency of stoppages in coalmining and in other industries before 1957 could probably be explained by similar factors – the most obvious being cyclical fluctuations in economic activity[2] – an explanation of the *declining* strike frequency in coalmining after 1957 has to be sought within coalmining itself.

There has been a marked contraction in the coalmining labour force since 1957. In part, then, the decline in coalmining strikes since

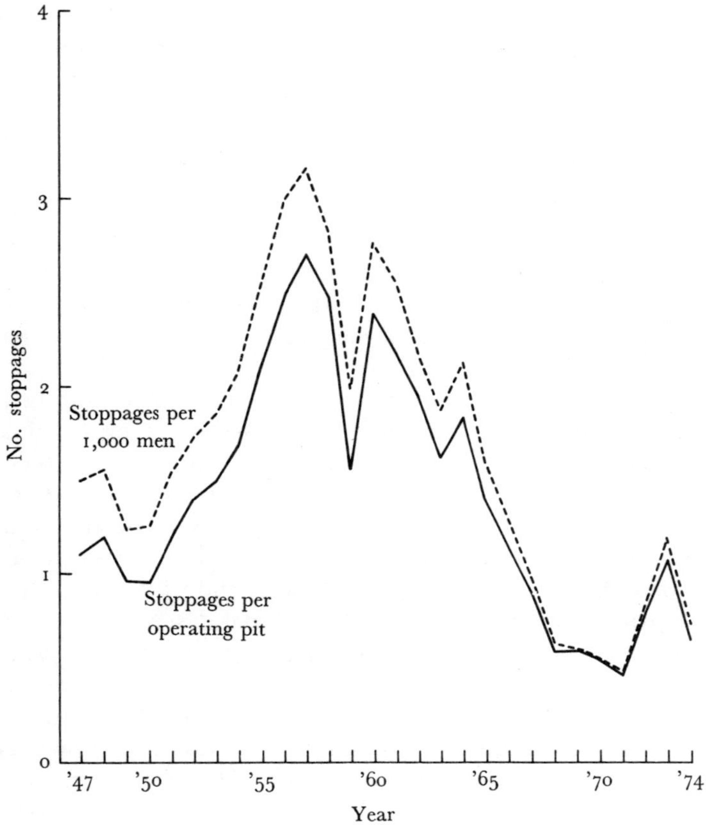

Figure 10.2 Work stoppages per 1,000 men employed and per operating pit in coalmining 1947–74

Source: Department of Energy, *Digest of UK Energy Statistics*

then might be a statistical phenomenon. These statistical effects are eliminated in Figure 10.2, which shows the number of recorded stoppages in coalmining as a ratio of both the number of wage earners employed by the National Coal Board and the number of operating pits in each year since 1946. These figures show that, although the percentage decline in the strike rate since 1957 is certainly smaller on these two statistical yardsticks than when no such adjustments are made, the decline in strike frequency is more than a statistical illusion brought about by the industry's contraction; stoppages per 1,000 men employed declined from 3.1 in 1957 to 0.57 in 1970, while the number per operating pit fell from 2.65 to 0.50. For example, if one takes the 1957 rate of 3.1 work stoppages per 1,000 men employed and applies it to the labour force of 283,000 in 1970, this would imply, *ceteris paribus*, a strike frequency of 877 in 1970, which is approximately 650 strikes more than were actually recorded in that year.

10.2 The change in wage payment systems hypothesis

Of the various occupational groups in coalmining, faceworkers have traditionally been the most strike prone. Baldwin,[3] for example, found that in 1949 70% of all strikes in the Yorkshire coalfield involved colliers and fillers, and the table he provided[4] showed that faceworkers as a whole accounted for over 80% of total work stoppages in that division that year. The percentage of strikes involving faceworkers may have been unusually high in Yorkshire. Even so, Baldwin's findings, and those of later students[5] of coalmining strikes, have tended to confirm that the largest proportion of strikes have involved direct production workers at the coalface. They have also suggested, like Baldwin, that the higher-than-average strike propensity of faceworkers was explained by the piecework method by which they were paid, the principal features of which we have described and analysed in Chapter 4. And the main source of discontent in this respect was associated with fluctuations in earnings produced by changes in seam conditions,[6] as well as by men having to transfer between seams of different paying capacities and managers withholding allowance payments for abnormal conditions as a form of disciplinary sanction.[7] Each of these sources of earnings variation could thus be related to the indeterminacy of earnings governed by the old price lists, which admitted scope for continual haggling between the faceworkers and their immediate supervisors.

We have described elsewhere[8] the changes which have occurred in wage-payment systems at the coalface since nationalisation. The most dramatic of these was, of course, in 1966, when fixed daywages gradually replaced piece rates on power-loading coalfaces. Whatever the effect of the NPLA and the 1971 revision of the elsewhere underground piecerate structure, these agreements hardly account for the substantial decline in work stoppages in earlier years: rather it is on the early district power-loading agreements that students have focussed their attention when trying to explain the earlier decline in strike frequency.

Thus Hugh Clegg[9] writes:

'The [National] Power Loading Agreement was the culmination of a long process of mechanisation and of adjustment in the traditional methods of wage-settlement. The proportion of coal output which was power loaded rose from 4% in 1950 to 70% in 1965. Over the period divisional power-loading agreements were negotiated coalfield by coalfield. Except for the Scottish agreement, which settled a time rate for power-loading ahead of the industry agreement, these divisional agreements contained two features. They retained a margin of pay fluctuating with output, but the major part of weekly pay was a fixed rate; and they removed the most important decisions on the pay of power-loading faceworkers from colliery bargaining to discussions between divisional coal boards and the area officials of the National Union of Mineworkers. Without detailed investigation of the effect of these agreements on the number of disputes, division by division, it is not possible to link the two with certainty, but it is at least plausible to suggest that the agreements led to a progressive reduction in the propensity of faceworkers to strike by diminishing the range of fluctuations in earnings and the scope for fragmented bargaining over pay in the colliery.'

Before we explore Clegg's hypothesis, one or two factual inaccuracies in the above quotation should be corrected. In the first place, the power-loading agreements were concluded on a wage district, not a divisional, basis except in Scotland, South Wales and Yorkshire, where the wage district and division were synonymous. Second, and more serious, is Clegg's general description of the agreements themselves. The Scottish agreement, for example, as we explained in Chapter 4, made provision for a completion-of-task bonus which, although fixed, was only to be paid if the set task was achieved.[10] This pattern was *also* followed in the Durham agreement. The difference between the agreements, however, was that, whereas in Durham wage *increases* were given on the completion-of-task bonus, they were given on the basic rate in Scotland. Thus the basic rate became a larger proportion of pay in Scotland than in Durham. As we explained in Chapter 4, though, all the other district power-loading agreements (except Yorkshire) were of a piecerate type and, although there was an agreed minimum (above the grade day rate) below which earnings could not fall, they did vary quite substantially with output and it was not unusual in some districts, particularly those of the East Midlands, for actual pay to exceed the minimum by 100%. In Yorkshire, power-loading teams were allowed to choose between *two* agreements: a piecework agreement and an enhanced daywage agreement of the Durham type. In Lancashire, however, the NUM refused to negotiate a district power-loading agreement at all, feeling that subordination to a higher level of wage negotiating authority would increase control over wages. There, the wages of power-loading operators were governed by local pit agreements. In

Kent, on the other hand, there was again no district power-loading agreement until 1965 – one year before the NPLA.

The role of area officials in pay determination under district piecework agreements is also overstated by Clegg. Wage norms were certainly negotiated at district level, but work targets and manning schedules were still subject to local negotiation and, as many areas refused method study, these norms were hardly as consistent between pits and coal faces as Clegg implies. Furthermore, there was still plenty of room for argument over such items as waiting time and make-up payments in so-called abnormal working conditions.

If, by lessening grievances over wage issues and wage fluctuations in particular, the growing importance of power loading in the late 1950s had a dampening effect on the industry's strike frequency, we might predict that this would manifest itself in a decreasing proportion of stoppages due to wage grievances, as well as an absolute decline in the overall strike frequency. Table 10.1, which gives the percentage allocation of coalmining strikes between the Department of Employment's principal cause headings, however, reveals no such tendency until after the NPLA in 1966. Indeed, the outstanding feature of Table 10.1 is the relative constancy of the percentage of work stoppages due to wage issues for the sub-periods between 1950 and 1965. And the fact that sub-periods following the NPLA *do* show a marked redistribution of strikes away from wages tends to confirm

Table 10.1. *Work stoppages in coalmining classified by principal causes 1946–73*[a]

Stated cause	Percentage of work stoppages						
(Dept. of Employment list headings)	1946–49	1950–53	1954–57	1958–61	1962–65	1966–69	1970–73
Wages	43.5	47.6	49.4	47.9	48.3	37.5	22.7
Employment	5.1	4.7	4.3	5.2	3.3	2.0	1.2
Work arrangements, rules, discipline *and* personnel	45.6	43.2	43.7	43.3	47.7	60.0	72.3
Others	5.8	4.5	2.6	3.6	0.7	0.5	3.8
	100.0	100.0	100.0	100.0	100.0	100.0	100.0

[a]The Causal Classification adopted by the Department of Employment has changed considerably since 1946. We have adopted general headings which embrace these changes and thus achieve greater consistency for the period as a whole. The distortions, if any, produced by our consolidation are unlikely to be large in the case of coalmining, where over 90% of stoppages come under the wages and 'working arrangements, rules and discipline' and 'personnel' headings in any case.

Source: Department of Employment, *Gazette*

the effect which we would have predicted if greater earnings stability had been a major cause of the industry's declining strike frequency *before* 1966.

One must, nevertheless, be very careful when dealing with statistics which relate to a whole industry, particularly in coalmining where miners have traditionally displayed a very uneven propensity to strike in different coalfields. Furthermore, district power-loading agreements varied a great deal, as did the dates at which power-loading agreements were concluded[11] – a fact, incidentally, which again makes one suspicious of the wage hypothesis as an explanation of a declining strike frequency which can be pinpointed so precisely to the late 1950s.

A regional analysis of coalmining strike trends necessitated changing our data source to keep the exercise to manageable proportions. The Department of Employment obtains details of stoppages of work (or 'disputes,' as they are called) from the NCB. These give the name and location of pits which have experienced stoppages but, because of changes in the Coal Board's regional boundaries, accurate allocation of the pits between wage districts is a difficult task, and exceptionally time-consuming given the number of stoppages in coalmining over the period which we have to cover. Therefore, the writer has relied on the NCB's source data, which were generously made available to him. These, however, differ from those published by the Department of Employment in several important respects.

First, the NCB dispute records include what it calls 'work restrictions', as well as actual stoppages. The former type of action is taken to refer to 'go-slows', which are notified to the various NCB industrial relations departments *when there is clearly an issue in dispute which has to be processed through the conciliation machinery*. It has been impossible to separate work restrictions and stoppages from the NCB's taped dispute records for the years 1947–60, so we have had to use a combined figure for these and, for the sake of comparability, subsequent years. Second, whereas the Department of Employment records exclude all stoppages of work which involve less than ten workers or lose less than 100 working days, the NCB records include such disputes. Third, whereas the Department of Employment attempts to classify a strike as a stoppage involving a single issue irrespective of the numbers of firms involved, the NCB classifies a stoppage (or a work restriction) according to the number of establishments (pits in this case) involved. Although it is highly probable that even the Department of Employment has some difficulty in deciding how many pits are involved in a stoppage over a particular issue (and may thus occasionally over-record the frequency of strikes in this

Table 10.2. *Stoppages and restrictions of work per 1,000 wage earners employed and per operating pit by NCB Divisions 1947–65 (annual averages)*

Division	1947–49	1950–53	1954–57	1958–61	1962–65
Scotland					
No. of disputes	475	768	1103	846	300
Disputes per 1,000 men	5.8	9.3	13.1	11.4	5.8
Disputes per pit	2.5	4.1	6.3	5.8	3.5
Northumberland and Durham					
No. of disputes	64	53	97	41	31
Disputes per 1,000 men	0.4	0.4	0.7	0.6	0.3
Disputes per pit	0.3	0.2	0.3	0.3	0.2
Yorkshire					
No. of disputes	393	493	911	647	489
Disputes per 1,000 men	2.8	3.6	6.6	5.2	4.4
Disputes per pit	3.8	4.3	8.2	6.0	4.8
East Midlands					
No. of disputes	44	41	50	53	51
Disputes per 1,000 men	0.5	0.4	0.5	0.6	0.6
Disputes per pit	0.5	0.5	0.6	0.6	0.6
West Midlands					
No. of disputes	26	47	92	94	74
Disputes per 1,000 men	0.4	0.8	1.6	2.0	2.0
Disputes per pit	0.4	0.8	1.7	1.9	2.1
South Wales					
No. of disputes	380	463	1056	671	630
Disputes per 1,000 men	3.3	4.3	10.1	7.5	8.8
Disputes per pit	1.6	2.6	6.6	5.0	6.5
North Western					
No. of disputes	167	111	142	139	85
Disputes per 1,000 men	2.8	1.9	2.5	3.0	2.3
Disputes per pit	2.2	1.6	2.0	2.3	2.1
Kent					
No. of disputes	12	6	24	18	10
Disputes per 1,000 men	1.9	1.0	3.4	2.9	1.9
Disputes per pit	2.9	1.5	5.9	4.5	2.5

Source: National Coal Board

industry as compared to others), on average the NCB will, for this reason alone, record more stoppages than will the Department of Employment. Unfortunately, it has been impossible to adjust the NCB data on a single issue basis. Our figures will thus be affected by the 'degree of inter-pit solidarity' displayed in particular areas.

Despite the discrepancies between the NCB data used in the subsequent analysis and those of the Department of Employment, all of which inflate the disputations recorded by the former compared with the latter, one feels that the NCB data do permit some meaningful conclusions in respect of the principal question to which we are seeking an answer here: namely, were the district power-loading agreements instrumental in the reduction in the miners' propensity to strike?

The incidence of work stoppages and restrictions has always varied quite markedly between the coalfields. Scotland, Yorkshire and South Wales, for example, which together have accounted for no more than 54% of the NCB's pits and provided employment for less than half the industry's wage earners, have been responsible for over 85% of work stoppages and restrictions recorded in most post-war years. This and some other aspects of dispute liability in the various coalfields are shown in Table 10.2, which gives for each NCB division the average number of stoppages and restrictions combined, as well as the combined stoppages and restrictions rate for one three-year, and four four-year periods, between 1947 and 1965. Our primary concern in the subsequent discussion, however, is to explain the observed trend, rather than the differences in dispute liability between the NCB divisions.

The mid 1950s represent something of a watershed in the post-war dispute history of *most* coalfields. The main difference, however (and one which suggests that, whatever factors were influencing the miners' dispute liability, their impact was not uniform), seems to be whether there was an actual fall in the dispute incidence in the late 1950s, or whether it was merely stabilised.

The most pronounced falls in the frequency of disputes and dispute rates in the late 1950s occurred in Scotland and Yorkshire, and, in the 1958–61 period at least, in South Wales. A reduction in dispute incidence was also experienced in Northumberland and Durham and the small Kent coalfield. The rate stabilised in the North Western and East Midlands divisions but, if anything, rose slightly in the West Midlands division.

Let us now turn again to the role which the district power-loading agreements played in these observed trends in the dispute liability of individual NCB divisions. And, as we are concerned especially with explaining the sudden decline in dispute frequency in the late 1950s, let us turn our attention to the coalfields of Scotland, Yorkshire and South Wales, which account for by far the largest part of the decline. Figure 10.3, which gives for each of these three divisions, over the period 1947 to 1966 (the year prior to the introduction of the NPLA),

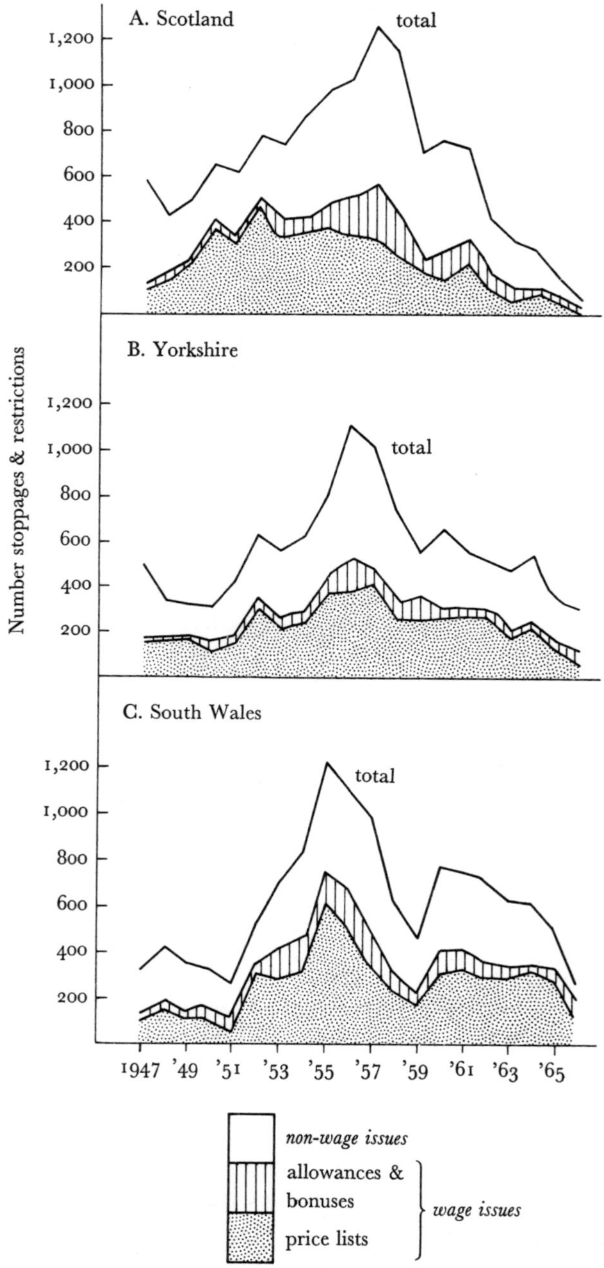

Figure 10.3 Work stoppages and restrictions in Scotland, Yorkshire, South Wales 1947–66
Source: NCB.

the overall year-to-year frequency of disputes, as well as the numbers attributable to wage disputes concerning price lists on the one hand and allowances and bonuses on the other, and non-wage issues, will help us in this respect.

The most interesting aspect of these figures is the close correspondence of the variations in the overall strike frequencies. True, the actual years in which the strike frequency turns down in the mid 1950s in the three coalfields does vary, this being 1956 in South Wales, 1957 in Yorkshire and 1958 in Scotland. But these variations are mainly accounted for by the incidence of particularly large solidarity disputes in the various coalfields.[12] The second point of interest, however, is the rather different trend in the incidence of disputes over the three main causes identified in our figures as between the divisions. We can examine this in more depth with the help of Table 10.3, which gives the percentage contribution of the three causes identified in Figure 10.3 – price lists, allowances and bonuses and non-wage issues – to total disputes in each division in various sub-periods between 1947 and 1965.

In Scotland, where the power-loading agreement fixed a daywage rate and completion-of-task bonus, disputes over price lists steadily declined as a proportion of the total as their number stabilized after 1954 and later declined. The Scottish power-loading agreement was

Table 10.3. *The percentage of work stoppages and restrictions due to principal causes (NCB)*

Division	1947–49	1950–53	1954–57	1958–61	1962–65
		(per cent – annual average)			
Scotland					
Price lists	30.5	49.2	33.2	25.0	29.4
Allowances and bonuses	3.5	4.4	12.7	12.7	14.3
Other	66.0	46.4	54.1	62.3	56.3
Yorkshire					
Price lists	40.2	38.4	39.1	40.7	40.1
Allowances and bonuses	3.3	6.9	9.5	8.0	3.0
Other	56.5	54.7	51.4	51.3	46.9
South Wales					
Price lists	31.5	43.6	43.0	42.1	52.6
Allowances and bonuses	9.2	10.1	12.6	7.2	3.5
Other	59.3	46.3	44.4	50.7	43.9

Source: National Coal Board.

introduced in 1954 and we feel justified in crediting this redistribution to the agreement. But the agreement did *not* prevent the total number of disputes from rising! It also seems to have encouraged Scottish faceworkers, who remained on the old price lists and did not benefit from the more stable power-loading wage, to focus increased attention, both absolutely and proportionately, on improving allowances and bonuses, thus to some extent offsetting the favourable effect of the power-loading agreement on the incidence of wage disputes.

A similar effect is discernible in Yorkshire following the introduction of that division's first power-loading agreement in 1951. Thus, the number and proportion of work stoppages and restrictions over allowances and bonuses increased during the period 1950–53 and until the late 1950s. In this division, though, the proportion of stoppages and restrictions over price lists remained fairly constant over the whole period 1947–65. So it was not until the early 1960s that the proportion of stoppages and restrictions over wage issues actually fell, which is hardly what one would have expected if the new-style power-loading agreement was causing the incidence of stoppages and restrictions to decline. The South Wales power-loading agreement was introduced in 1957. This seems to have had an opposite effect to that of Scotland and Yorkshire, for the proportion of disputes due to allowances and bonuses *declined* while those due to price lists remained constant. Whatever effect the South Wales agreement had on the actual incidence of disputes, however, seems to have been of short duration, for by the early 1960s the proportion over *wages* was higher than the level of any previous post-war period.

It would thus appear that the impact of the new style district power-loading agreements on the industry's strike frequency was far from being as positive or uniform as some have hypothesised. The Scottish agreement *does* seem to have reduced the miners' propensity to strike over wage issues and probably served to keep that division's strike rate below what it would otherwise have risen to. The same conclusion might also apply to the Durham agreement, which was similar to that of Scotland. In the other areas, the impact of the new-style agreements is difficult to ascertain. At best, they seem, in the longer run, to have reduced the number and proportion of grievances over wage issues *other* than price list questions, but this could be as much due to the more coordinated approach to allowances and bonus payments in most areas during the early 1960s as it was to the power-loading agreements as such.

It could be argued, of course, that by taking some of the physical effort out of facework, the power-loading machines themselves brought about a reduction in the miners' strike frequency in the late

1950s. But if so, why did this frequency not occur earlier? In Yorkshire, for example, power loading already accounted for 25% of output by 1958. We would suggest, therefore, that the new-style power-loading agreements only partly explain the declining propensity of miners to strike in the late 1950s and that such positive effects as they had were probably confined to the Scottish coalfield, with Yorkshire as a marginal case. Even in these coalfields, however, we suspect that other influences were at work to reduce the overall strike frequency. We shall return to this question shortly.

10.3 The National Power Loading Agreement

The National Power Loading Agreement of 1966 took the process of wage reform at the coalface further than any of the district agreements. The NPLA established a fixed daywage rate for each coalfield and, although the district rates differed at first, they were eventually brought into equality. That agreement was followed in 1971 by the replacement of piece rates for daywage rates for tasks behind the coalface.

The number of strikes reported in coalmining continued to decline until 1972, although at a slower rate than during the period 1962–65. Even so, the largest part of that decline was attributable to a sharp reduction in the number of strikes over wage issues and we feel fairly confident that the NPLA had a favourable effect on the industry's strike *frequency* in the short run.

Although the bulk of our analysis in this chapter has concentrated on the declining frequency of strikes in coalmining following the mid 1950s, however, we can hardly ignore the fact that the strike frequency rose in the early 1970s. This upswing was accounted for almost entirely by an increase in the frequency of strikes over non-wage issues. An examination of both Department of Employment and NCB data revealed it to be attributable mainly to a greater frequency of work stoppages over refusals to accept what management regarded as reasonable alternative work and disputes over manning and work organisation. This latter tendency was, as we have already suggested, explicable by two factors. First, the NPLA and the Third Daywage Structure stabilised and equalised earnings for tasks which many felt were unequal in terms of the effort input required (the equalisation of wage rates between power-loading operatives and rippers is but one example). This undoubtedly made men more reluctant to do the more arduous tasks, despite clauses in the agreements which were to guarantee the NCB complete flexibility in the allocation of men between the tasks covered. The second factor, related to the first, was that both agreements shifted the emphasis from haggling over local wage additions to arguments over manpower allocation. In other words, there was a change of emphasis in *local* bargaining, from the wage, which was now determined nationally, to the effort input required for the given wage.

10.4 Strikes and the bargaining environment

We have already suggested that we suspect that factors other than changes in the system of coalface wage payment were operating to reduce, or in some coalfields stabilise, strike frequency in the late 1950s. And we suspect that it was more than mere coincidence that *in general* the initial downturn coincided with a slump in coal demand. This suggests that it may have been partly a reaction to what the miners saw as a weakened bargaining position in the face of a potentially falling demand for their services. An opposite response in the face of an increasing, but suppressed, demand for coal in the early 1950s would also explain why the strike frequency in the industry increased so sharply during that period, despite the spread of power loading. In this context it is also interesting to observe, in Figure 10.4, that the small rise in strike frequency in 1960 coincided with what proved to be only a temporary upswing in consumption and that the strengthening of the demand for coal in the early 1970s was also associated with an upturn in strike frequency.[13]

In addition to the annual inland coal consumption and the work stoppage rate, Figure 10.4 gives the number of issues per 1,000 employees which went through the pit dispute procedure. If the wage reforms had really been instrumental in reducing miners' grievances, we might expect this to have been reflected in fewer issues raised in procedure as well as in fewer stoppages. This certainly seems to have been the case during the two years following the introduction of the NPLA in 1966, which confirms our previous impression that this agreement did lessen discontent over wage structure issues in the short run. A similar tendency is noticeable between 1958 and 1959 when the work stoppage rate first declined. But, whereas, after a brief revival of strike activity in 1960, the strike rate fell quite sharply in 1964, the issues-in-dispute rate increased to an all time peak in 1963. In other words, the reduction in the strike rate reflected a reduced *willingness* of miners to strike, rather than a reduction in the number of grievances. It would appear, then, that after 1957 miners were more willing to resort to the conciliation procedure, as their bargaining power decreased. This again suggests the conclusion that the unfavourable economic environment of the late 1950s and 1960s was

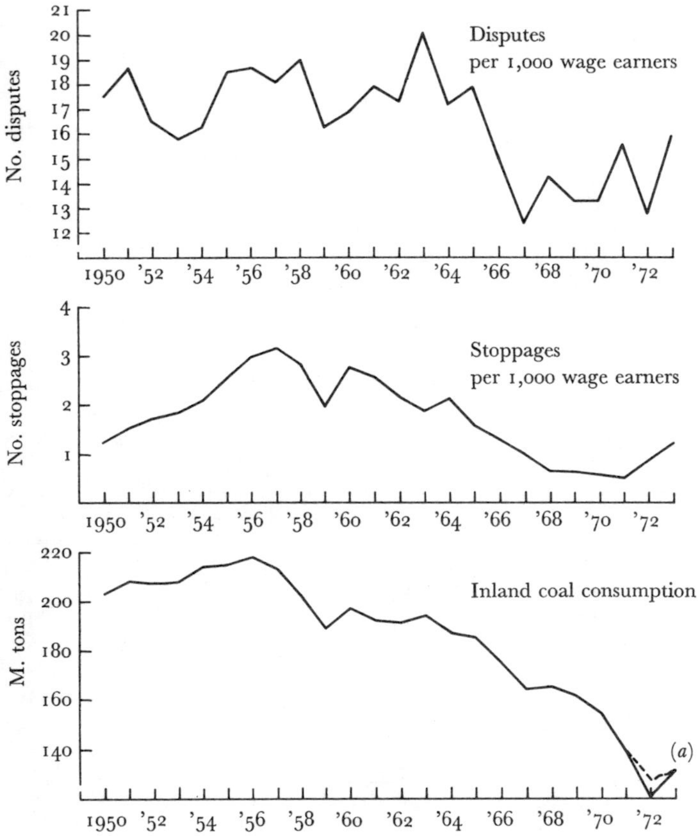

Figure 10.4 Disputes, work stoppages and coal consumption 1950–73.
[a]In 1972, inland coal consumption was affected by the national miners' strike. The hatched line corrects for this.
Source: NCB, Dispute Returns.

a major influence on what, in general, was a declining strike frequency in the industry.

10.5 Other dimensions of strike activity and a conclusion

Thus far we have deliberately refrained from referring to other dimensions of coalmining strike activity, such as the number of working days lost through stoppages and the number of workers involved in them. The reason for this is that, as we can see from Table 10.4, such interesting trends as these statistics reveal are of fairly recent origin: we refer in this context, of course, to the re-emergence

Table 10.4. *Some further dimensions of coalmining strikes 1947–74*

Period	Strikes	Working days lost (000s) (annual averages)	Workers involved (000s)
1947–49	1,014	710	248
1950–53	1,115	459	180
1954–57	1,887	649	266
1958–61	1,598	511	231
1962–65	995	337	149
1966–69	338	454 (206)	67 (36)
1970–73	205	2,250 (66)	133 (30)
1974	183	5,625 (58)	306 (n.a.)

	Working days lost per 1,000 men	Workers involved per 1,000 men	Working days lost per strike	No. workers involved per strike
	(annual averages)			
1947–49	990	346	700	245
1950–53	647	253	412	161
1954–57	919	377	344	141
1958–61	824	368	320	145
1962–65	668	292	338	150
1966–69	1,220 (554)	180 (97)	1,343 (609)	198 (106)
1970–73	8,212 (241)	485 (109)	10,975 (325)	649 (148)
1974	22,865 (235)	1,244 (n.a.)	30,737 (319)	1,672 (n.a.)

Source: *Department of Employment Gazette*

of the large solidarity strike, which was so much a feature of coalmining industrial relations in the early twentieth century.

After the first three years of nationalisation, during which there had been one or two relatively large stoppages – mainly over inter-union conflicts – the trend of working days lost and numbers of men involved in work stoppages tended to follow that of strike frequency until the late 1960s. This was because the typical coalmining strike was a very small, short-lived affair, much more so than our *average* statistics of working days lost per stoppage and numbers of workers involved would indicate. In 1965, for example, 58% of stoppages reported to the NCB involved less than 30 workers and 68% lasted for one shift or less.[14] In the late 1960s, however, there was a tendency for strikes in coalmining to become longer. In 1969, for example, there was a widespread unofficial stoppage involving 121,000 men, which lost nearly a million working days. This was followed in 1970 by another large stoppage in support of a claim for an extra £5 a week. This, too, was unofficial, although some members of the NUM executive were seeking to make it official. In 1972, the miners held their first national strike since the 1926 strike cum lockout, which cost nearly 11 million working days. This action was repeated in 1974 at the cost of over five million working days in mining and a three day week for large sections of British industry. These large strikes are excluded in the bracketed figures in Table 10.4.

The last three stoppages mentioned above are obviously the main factors in the sudden increase in working days lost and worker involvement in stoppages during the period 1970–74. But even when we exclude the 1969 stoppage from our data for the period 1966–69, one can still detect an underlying trend towards rather longer stoppages. Thus, whereas the annual stoppage frequency fell by over two thirds between 1962–65 and 1966–69, the number of working days lost (excluding the 1969 action) only fell by approximately one third. This raised the adjusted figure of working day lost per strike well above that for any previous period except 1947–49. As the adjusted figure for workers involved per strike fell during the period 1966–69, we can safely conclude that there was an underlying tendency for stoppages to become longer. This alone might have been a fair warning of the change in the climate of industrial relations which became so clearly manifest in the 1970s.

The emergence of the larger and longer (national) strike since the late 1960s was not, of course, peculiar to coalmining, although the miners could certainly be thought of as trend setters, in what eventually became much more concerted industrial action on basic pay issues in the public sector as a whole. And it is somewhat ironical that having heralded what was to have been a new era of industrial relations, following the complete abolition of piecework in the pits, the industry should return to such prominence in the annals of work stoppages. Our analysis in earlier chapters, however, suggests some reasons for this. Of these factors, four deserve special mention, namely the general trend of pay during the previous decade, the wage structure reforms themselves, a change in the NUM leadership and the changing market position of coal.

We saw in Chapter 8 that, in terms of average weekly earnings, the *money* wage relativity between miners and male manufacturing workers declined dramatically between 1957 and 1969. The implications of this declining wage relativity, however, were particularly onerous in net real income terms: we illustrated in Chapter 5, for example, that, having fallen quite dramatically in 1958, average net real incomes in mining had failed to return to their 1957 level by 1967 and by 1968 were again on the downturn and remained so until the 1970 wage settlement. Thus, whereas in 1957 the net real income of our average miner with a wife and two children was approximately 22% above that of his male counterpart in manufacturing, it was approximately 2.0% below the manufacturing figure in 1969. If, as has been suggested by some writers,[15] the increase in industrial militancy in the late 1960s more generally can be associated with a slowing down in the average rate of net real income growth, it is hardly surprising, in the light of net real income trends in mining, that coal miners should have been at the forefront of wage militancy.

If one is to explain the rather sudden change in miners' militancy in the late 1960s, however, this relatively adverse trend in their net real income must also be seen in the context of wage-structure trends following the 1966 NPLA. It will be remembered from Chapter 8 that wage equality among power-loading workers following the 1966 NPLA was to be achieved by raising the shift rates of faceworkers in what had previously been the lower-paying wage districts, such as Durham, Scotland and South Wales, faster than those of workers in the previously higher-paying districts, such as the East Midlands. As this was to occur within an overall pace of wage increase which was less than that in manufacturing industries as a whole, it meant that NPLA workers in the higher-paying areas suffered a more than average decline in net real incomes between 1966 and 1969. Thus the differential rates of income growth implicit in the trend to wage equality aggravated faceworkers in the traditionally less militant coalfields of the East Midlands.

The wage reforms also help to explain, at least partly, the greater solidarity among coalfields and occupational groups. The abolition of piecework and the wage equality which came with it gave faceworkers a hitherto unaccustomed unity. Also, by transferring power from the pit and district to the national negotiating table, it put the onus for improvements in their living standards fairly and squarely on the shoulders of the union's national negotiators, and established a direct negotiating link between the previous pieceworkers and their traditionally daywage colleagues. For the first time in its history, then, the National Union of Mineworkers had become a national union in more than name alone.[16]

The third factor in explaining the more coordinated and determined approach of the NUM was undoubtedly the change in its leadership, and in its attitude to wage questions. For a decade after 1957, the NUM was as much, if not more, concerned with preserving employment in the industry as it was with improving wages; its leaders having been convinced of a probable trade-off between the two. A distinct shift in emphasis was apparent in the late 1960s, however, with first the retirement of Will Paynter as General Secretary and his replacement by Lawrence Daly, followed in quick succession by the election of Joe Gormley to replace Sir Sid Ford as President and the replacement of Sid Schofield by Mick McGahey as Vice President. Lawrence Daly, in particular, had been sceptical about trading wages for jobs and had been a strong advocate of treating wages and employment as separate negotiating issues. Later he was undoubtedly supported in this matter by his Vice President, with whom he had worked closely in Scotland where, in their capacity as Area Secretary and President respectively, they had fought relatively successfully on the wages front and were renowned also for taking realistic views on pit closures where they were deemed necessary.

The most obvious sign of a change in NUM thinking came in 1970, when the executive rejected an NCB offer which fell within the Tory government's informal pay increase norm of 12% per annum in the public sector, and against the background of fairly widespread unofficial action sought the support of its membership through a national ballot to pursue a much larger claim.[17] Although the NUM executive did not obtain the requisite two thirds support for official action, the 1970 negotiations were a landmark in post-war mining industrial relations in that the NUM had actually sought a national mandate for strike action. At the union's annual conference the following year, the majority necessary for the executive to call for national action was reduced to 55% and this proved more than sufficient to ensure support for the first nationally supported strike the following year.

The fourth factor conducive to increased militancy among coal-miners in the early 1970's was of course the change in the market environment of coal and manpower trends, which were moving in a direction favourable to the miners even before the increase in oil prices. While the Coal Board was operating with a current labour surplus in the late 1960s and early 1970s, there was evidence[18] that the industry would soon experience manpower shortages if planned production levels were to be met and particularly if the level of unemployment were to fall. Thus, rather than the NUM having to worry about the effect of marginal pit closures on employment (and its own membership), it was soon to be the turn of the NCB to worry about labour shortages in its most productive pits. This argument was not exploited to the full until the 1974 dispute, but the trends no doubt lessened any anxiety the NUM may have had about the unfavourable effects of wage increases on employment.

Insofar as we are attempting to explain the changed character of industrial action in coalmining in the 1970s, however, we regard the changes which occurred in coal's market position as of marginal importance compared with the impact of past wage decisions and trends. Indeed, we think it highly likely that miners would have taken a more determined and concerted stand on wage issues, irrespective of the shift in bargaining power in their favour, although this is now a key factor. And it could be that the most significant consequence of the industry's revised wage structure was its unifying effect on the miners' union, rather than its effect on the *frequency* of strikes. Certainly, the strike pattern to have emerged in the early 1970s must be a salutary warning to anyone who believes that procedural reforms in collective bargaining are sufficient, by themselves, to mitigate overt signs of industrial unrest.

11. WAGE RELATIVITIES AND LABOUR MOBILITY

The decline of coal demand, combined with technological change, produced a substantial contraction in the coalmining industry's labour requirement after the mid 1950s: between 1957 and 1973 colliery manpower, having been relatively stable between 1948 and 1956, fell from 704,000 to 268,000. The renewed emphasis on coal, following the sharp increase in oil prices in the early 1970s, however, made it imperative that coalmining employment be prevented from falling further if the industry was to meet its new output commitments. The sudden change in the market environment greatly strengthened the NUM's hand in wage negotiations and, in March 1974, after a bitter, but successful, confrontation with the Conservative government, the miners achieved a substantial improvement in their pay position relative to other workers.

The NUM's pay claim, which was given special consideration under the government's incomes policy, was the object of fairly rigorous, if rapid, examination by the Pay Board.[1] The terms of the settlement agreed by the NUM and the new Labour government elected in March did not follow the Pay Board's recommendations in every detail. Nevertheless, they did reflect the Pay Board's opinion that an increase in the miners' wage relativity was an essential part of any package which sought to halt the recent decline in the industry's labour force. As the role of relative wages in manpower distribution has caused considerable controversy among labour economists, the substance of the Pay Board's argument is clearly of more than practical interest. In this chapter, therefore, we focus on some aspects of labour supply in coalmining and examine its sensitivity to relative wages. After a brief résumé of earlier work on the role of wages in manpower distribution, the chapter is divided into three sections which consider, in turn, the direction and determination of labour flows to and from coalmining, manpower trends in coalmining in relation to output targets in the 1970s, and the effects on coalmining employment of considerably raising the miners' external pay relativity in the mid 1970s.

11.1 Wages and manpower supply

Views on the role of relative wages in manpower distribution differ quite markedly between economists. On the one hand there is the traditional market view, which suggests that relative wages are important in manpower allocation. On the other hand, there is the view that manpower allocation is determined mainly by employment opportunities and the intensity with which employers recruit labour, and that relative wage adjustments are unnecessary to secure alterations in the allocation of manpower.

An early attempt to test the market and job opportunities theories of labour movements was made in 1959 by Professor Brian Reddaway,[2] in response to the Cohen Council's[3] advocacy of a flexible approach to wage structure on the grounds of manpower allocation. Professor Reddaway's pioneering study, together with later, supporting, findings from an OECD[4] study, constitutes the main evidence in support of the job opportunities theory, despite his own warnings regarding data limitations and problems of interpretation.

Reddaway's study was designed to examine the extent to which industries expanding or contracting their labour forces gave increases in average hourly earnings which were higher or lower relative to the average, or ended up by paying more or less than the average, over the period April 1951 to April 1956. It was Reddaway's view that the market theory could be supported if there was a strong positive correlation between the wage and employment variables. Like most recent students of the wage structure, he found that, in comparison with employment changes, variations in the growth of average hourly earnings between industries were fairly small. When he correlated the two sets of data for the fourteen manufacturing orders he did find a strong association between rates of wage and employment change. The same test on the 111 industries from which the orders are constituted, however, produced only a weak association. Furthermore, there was no association within particular industrial orders, wage increases there being fairly uniform while employment changes were very varied.

Although the close association of wage and employment changes between industrial orders supported Reddaway's interpretation of the market hypothesis, he used the much weaker associations from the intra-order classification to conclude that changes in relative wages are unnecessary to bring about employment changes. It would seem, though, that Reddaway's preference for this interpretation was partly due to his feeling that a wages policy which admitted alleged manpower deficiencies as an argument for exceptional wage consideration would tend, through the forces of fair comparison, to exacerbate inflation.

An OECD study, based on similar data, endorsed Reddaway's findings, concluding that 'there is no evidence of a strong systematic relationship between changes in earnings among individual industries and variations in relative employment'.[5]

To be fair to both Reddaway and the OECD, neither said explicitly that wages were wholly unimportant in influencing the direction of labour movements; neither denied that an increase in relative wages might make the task of recruiting workers easier. Rather, the important point of the job opportunities hypothesis was, as explained by Reddaway,[6] 'that the *main* way in which employment will be either raised or reduced is through "direct action" by employers, and that only exceptionally will they have to include a change in the relative wage offered in order to secure the desired number of workers – or at least that the change in relative wages will usually be so small that it is hard to detect in a world where other wages are also changing'. The OECD[7] study, however, did point to one exception where wages might be important in manpower distribution. This was the case of an industry paying earnings substantially less than the average or whose relative wages had shown a considerable decline. These industries, the report postulated, might experience recruitment difficulties.

The studies by Reddaway and the OECD are open to a number of criticisms. First, the so-called market theory of wages and labour allocation in no way implies an association between relative rates of wage *advance* and net employment changes. At any moment a firm, or industry, may already be paying a wage rate which is quite sufficient for it to obtain all the labour it requires, or even one which is above that necessary to get the labour required, as it will inevitably take time for relative wages to adjust to changing employment needs and vice versa.[8] Thus, relative wage *changes* in *any* direction could be quite consistent with net employment changes in any direction although a labour force is likely to be in a strong bargaining position if employment needs are expanding rather than contracting. Second, and especially important, a net change in employment tells us virtually nothing about the ease with which it was achieved, or the relative stability of the labour force as reflected in rates of labour turnover or, as a consequence, about changes which may have occurred in the character of the labour force.

If there is high unemployment among people with the skills which a firm wishes to employ in larger numbers, we may not expect the employer to feel obliged to raise relative wages to obtain them. If he has to attract those skills in a tight labour market, however, he might well attempt to pay more relative to other employers. If he succeeds in obtaining the additional labour at the expense of other employers, he may get a net addition to his labour force and the other employers will suffer a net reduction. It may take time for other employers to adjust to their net loss, but when they do realise their plight, they too may seek to raise wages or be put under more pressure to do so, if the labour force is organised. Over, say, the next year, the wage relativity between the employers may be restored or even move in the other direction. The statistical correlation over (say) a two- or three-year period between wage relativities and employment changes, however, could be positive or negative, depending entirely on whether changes in wage relativities and employment changes correspond in timing.

In any case, the statistic of net employment change only tells us what happened to employment: it does not tell us how much labour a particular employer wanted – which may itself depend on the wage rate – or how deficient his labour force was at any moment in time; and it is those considerations which will help to determine his attitude to wage relativities. If it seems reasonable to postulate that an employer's wage behaviour (as well as an employee's wage sensitivity) will depend, in part, on the supply of that skill in relation to the demand for it at a particular time, it is equally unreasonable to assume that this postulate only holds true for industries which have an *increasing* demand for labour. In a tight labour market, an industry with a falling employment trend may find it just as difficult, if not even more so, to retain labour or replace it, as one which is trying to increase its employment. Thus, in the absence of detailed analysis of the determinants of labour supply and demand, virtually any observed wage structure trend is consistent with any observed trend in the employment structure; particularly when the problem of time lags is brought into the picture.

It follows from our discussion so far that if one is adequately to test for the sensitivity of movements of labour to relative wages, it is imperative to focus attention on the labour flows from which net changes in employment (if they occur) are compounded. Again, though, we might expect some of these flows to be more responsive than others to relative wages and for employers to react more strongly to some than others.

An employer will have an employment deficiency if he loses more labour than he anticipated or fails to recruit sufficient labour to fill his job vacancies. The recruitment statistic by itself may not be sensitive to relative wages, as it will almost certainly be influenced by the number of vacancies available. However, recruitment may be equal to, or less than, that required and this may influence the employer's attitude to wages. Unless one knows the employer's overall labour requirement, though, the recruitment figure will tell us little about the employment problem confronting him; nor will it tell us a great deal about employee reaction to relative wages. A more meaningful statistic here would be the number of applications an employer receives in response to advertised vacancies, but this information is virtually impossible to obtain. Similar problems occur when examining wastage, in that part of the total will be the product of non-economic factors, such as death retirement and dismissal,[9] which have nothing to do with relative wages. It is imperative, therefore, that we distinguish that part of wastage which is involuntary from that due to the worker's own initiative, which we can designate as voluntary wastage.

Insofar as the wages side of the equation is concerned, it is doubtful if the percentage *increase* in earnings is the most relevant statistic to adopt when looking at employment shifts or the turnover rates from which they are compounded. Insofar as a worker's job choice may be influenced by earnings, it is most likely to be the actual differential between occupations which will figure in his calculations, rather than the movement in that differential. Nor is it necessarily average hourly earnings which are the relevant earnings statistic. A worker will usually want to know how much he can earn in a week. The hourly wage rate is important in this respect, but so is the number of hours he can expect to work. If the difference in hourly earnings between two occupations is marginal, the worker may very well accept the job at the lower hourly wage rate if it offers him the opportunity of working longer hours, and thus of gaining higher weekly earnings.[10]

While one can clearly take issue with the theoretical model underlying some of the work on wages and employment changes, many of the difficulties outlined above undoubtedly arise from the absence of information on labour flows and, at the time some of the research was conducted, appropriate wage data. The mere fact that more appropriate data are available for the coalmining industry thus makes a study of mining employment and relative wages pertinent to labour market analysis more generally. Indeed, as most of the nationalised industries regularly publish data on labour recruitment and wastage, it is surprising, given that the proof of so much labour market theory depends on an analysis of labour flows rather than employment stocks, that labour economists have paid it so little attention to date.

11.2 The coalmining case

Employment in coalmining has changed very rapidly since nationalisation, as has the miners' wage relativity to manual workers in other industries. During the first decade of nationalisation, the industry was preoccupied with maintaining employment. The reduced demand for coal, combined with a strongly rising productivity trend after 1957, served to reduce greatly the industry's manpower requirements. The NCB reduced labour recruitment and relied on wastage, both involuntary and voluntary, to achieve the necessary employment cuts against what was a substantial fall in the miners' wage relativity. What role, therefore, did relative wage trends in coalmining play in manpower policy? We shall examine this problem by first examining the determinants of the voluntary wastage rate in coalmining.

Our model is very simple, in that it postulates voluntary wastage from coalmining, which we define as the number of miners leaving the industry's employment each year for reasons other than dismissal, death and redundancy, as a percentage of mid-year employment, to be related to the overall demand for labour (and thus indirectly the availability of alternative work) and the relative wage position of miners *vis-à-vis* workers in other industries who might recruit from the mines. For the first variable we have chosen the general unemployment percentage.[11] The choice of an appropriate wage relativity variable poses much greater problems but we can get some guidance in making our selection from Table 11.1, which shows the relative importance of different sectors as recruiters and suppliers of mining labour between 1959–67.[12]

The industrial direction of wastage from the mines and the sources of recruitment are fairly similar. If we look at recruitment first, we can see that, of those recruited from other industries, 39% came from manufacturing – the biggest recruitment from one sector being from bricks, pottery, glass and cement – 22% from construction, and 32% from industry orders 18–24, the most important here being transport and the distributive trades. The outflow of labour from coalmining showed a larger bias towards manufacturing employment (45.8%) than did the inflow, while the percentage going into construction was similar to the percentage recruited from that sector.

In the absence of an occupational breakdown, we cannot say a great deal about the skill mix of the flows. The importance of the construction industry as a source of mining recruitment and recipient of wastage, however, is hardly surprising, given the similarity of work and the wide geographical dispersion of construction, and we might expect the industry to be a major recipient of mining faceworkers.

Table 11.1. *Labour flows to and from coalmining: total, 1959–67*

1958 SIC order	Coal recruitment from No. (000s)	%	Coal wastage to No. (000s)	%
Agriculture, forestry, fishing (1)	11.8	6.7	6.6	2.0
Manufacturing total (3–16) of which:	69.2	39.1	152.9	45.8
bricks, pottery, glass, cement (13)	(14.2)	(8.0)	(30.3)	(9.1)
Construction (17)	38.9	22.0	70.6	21.2
Other total (18–24) of which:	57.1	32.2	103.7	31.1
transport & communications (19)	(19.6)	(11.1)	(25.2)	(7.5)
distributive trades (20)	(18.2)	(10.3)	(23.7)	(7.1)
Total	177.0	100	333.8	100

Source: Department of Employment.

Other mining skills are common to most of the other industries and, on *a priori* reasoning, we would expect craftsmen and surface workers to have a fairly wide industrial choice. Again, though, they will tend to go to industries which are concentrated in the region of the pits. This accounts for the high concentration of wastage to particular industries, such as bricks, pottery, glass and cement, for example.

The information in Table 11.1 suggests that, although they will necessarily be something of an abstraction, the appropriate wage relativities are those between miners and male workers in manufacturing and male workers in construction. As we would expect the voluntary wastage rate to be negatively correlated with both the level of unemployment and the relevant wage relativity, our two equations will be:

$$VWM_t = a + bU_t + c(RM)_t \qquad (1)$$

and

$$VWM_t = a + bU_t + c(RC)_t \qquad (2)$$

where

VWM = Annual voluntary wastage as a percentage of the coalmining labour force

U = Annual general unemployment percentage

RM = Annual wage relativity between miners' average weekly earnings and male manufacturing workers' average weekly earnings

RC = Annual wage relativity between miners' average weekly earnings and construction workers' average weekly earnings
a = a constant
t = year

Before 1952, miners had to obtain special permission from the then Ministry of Labour to leave the mines. This might be expected to reduce the sensitivity of the voluntary wastage percentage to our labour market variables, so we have selected 1952 as the base year for our analysis, although including earlier years did not substantially affect the results which are given in Table 11.2 for the period 1952–74 (with standard errors in parentheses).

These equations, estimated by least squares, tell us that year-to-year fluctuations in voluntary wastage from the mines are negatively related to the level of unemployment on the one hand and to the wage relativity between either miners and male workers in manufacturing, or miners and construction workers, on the other. In terms of explanatory power, however, the best results are given by equation (3), which combines the mining/manufacturing wage relativity with the general level of unemployment, as opposed to the combination of unemployment and the mining/construction wage relativity. The former equation explains 80% of the variation in the voluntary wastage rate compared with 72% in the latter. The effect of a one percentage point change in unemployment on the miners' voluntary wastage percentage is approximately eleven times that for a one percentage point change in the mining/manufacturing wage relativity, or a fifteen percentage point change in the mining/construction wage relativity. We did estimate an equation which incorporated unemployment and both wage relativity variables. The construction coefficient had the right sign but was statistically insignificant, reflecting the high collinearity between the two wage-relativity variables. In view of the evidence on labour flows, it probably makes most economic sense to specify the effects of the wage-relativity variables separately.[13] We can, however, restrict further analysis to the equation which combines the level of unemployment with the mining/manufacturing relativity as an explanation of variations in miners' voluntary wastage.

Figure 11.1 shows the expected values of voluntary wastage derived from equation (3) which incorporates the mining/manufacturing relativity, against the actual values over the period 1952–74. The relative importance of unemployment and the wage relativity to the voluntary wastage percentage is shown in Figure 11.2. In the latter figure, the vertical axis is *negative*, indicating that an increase in either the unemployment or wage relativity percentage *reduces* voluntary wastage.

Table 11.2. *Regression equations for annual voluntary wastage from coalmining between 1952 and 1974*

a	U	RM	RC	R^2	DW^a	
13.02	−2.20	−0.19		0.80	1.35	(3)
(0.79)	(0.31)	(0.02)				
13.49	−2.28		−0.15	0.72	1.42	(4)
(1.01)	(0.37)		(0.02)			

[a]The equations were also estimated with a correction for auto-correlation, but the results were not significantly different from those in the table.

Apart from the rather large divergence between the expected and actual voluntary wastage percentages in the early 1960s, explained perhaps by the rephasing of colliery closures, the equation appears to give a remarkably good estimate of voluntary wastage, particularly for the 1970s. The closeness of fit in the 1970s is particularly interesting, for it was during this period that the largest changes occurred, first in unemployment and then in the wage relativity. This is especially important when evaluating the 1974 wage settlement, which, taking the October figures, had the effect of raising miners' wages from 102% to 118% of those of their male counterparts in manufacturing, while unemployment was constant at 2.6%. Despite the enormous change in wage relativity, which one might expect to pose some problems for accurate prediction, the reduction in the voluntary wastage percentage between these two years was almost identical to that predicted by equation (3).

Perhaps the ultimate test of the accuracy and stability of the model as an estimating device, however, is provided in Figure 11.3, which, on the basis of a regression for the period 1952–70, shows the expected voluntary wastage percentage for the same period and the voluntary wastage forecast for 1971–74 using that equation, compared with the actual voluntary wastage percentages. Figure 11.4 gives revised estimates of the contributions which unemployment and the wage relativity would then have made to the voluntary wastage percentage. The coefficients upon which these estimates are based are given below in equation (5).[14]

a	U_t	RM	R^2	DW	
13.88	−2.69	−0.20	0.77	1.20	(5)
(1.25)	(0.58)	(0.03)			

The value of the coefficients for unemployment and the wage relativity in equation (5) are so similar to those of equation (3) that it is hardly surprising to find the two equations giving fairly similar predicted values of voluntary wastage for the period 1971–74, or that equation

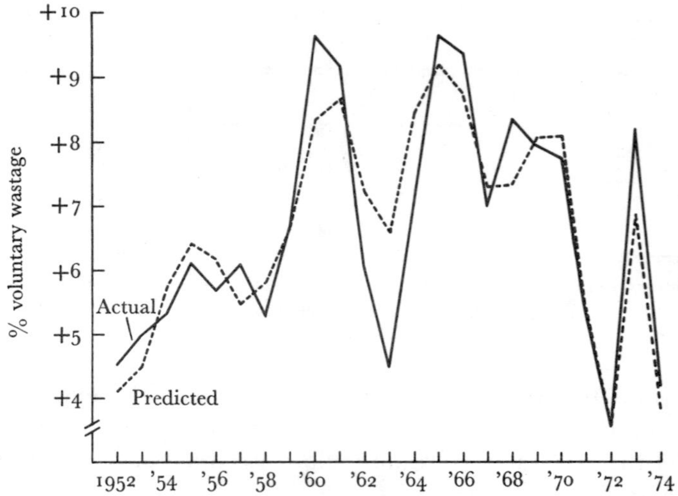

Figure 11.1 Actual and predicted voluntary wastage 1952–74
Source: NCB.

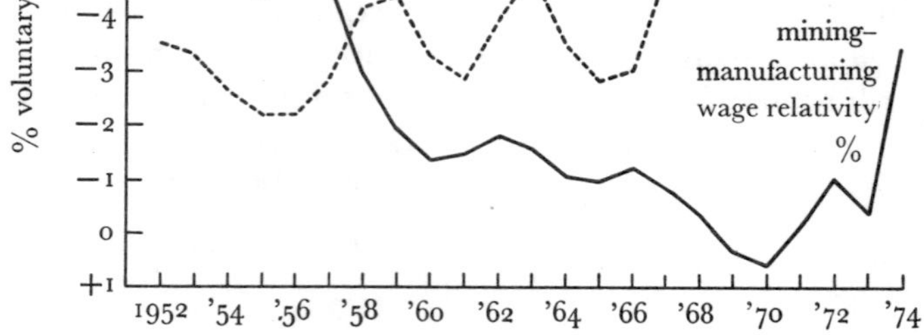

Figure 11.2 The contribution of unemployment and the mining/manufacturing earnings relativity to mining voluntary wastage 1952–74
Source: Department of Employment.

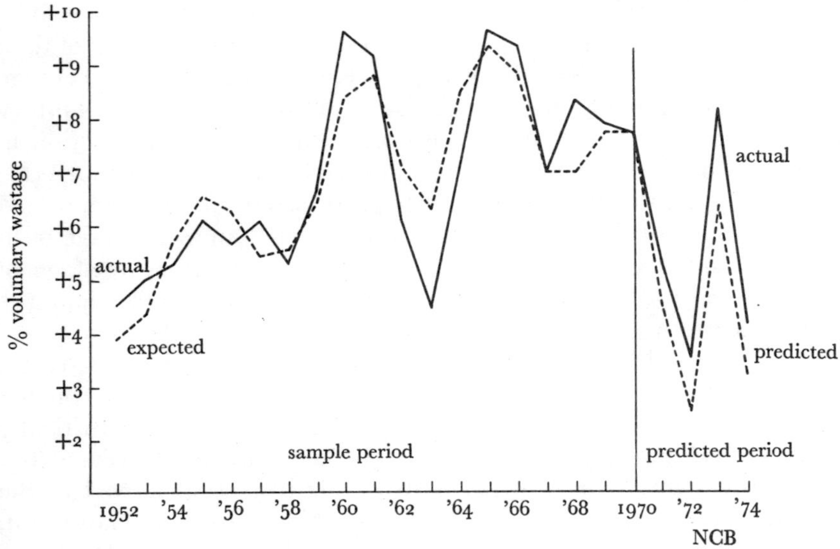

Figure 11.3 Actual and predicted voluntary wastage 1952–74, from 1952–70 relationship
Source: NCB.

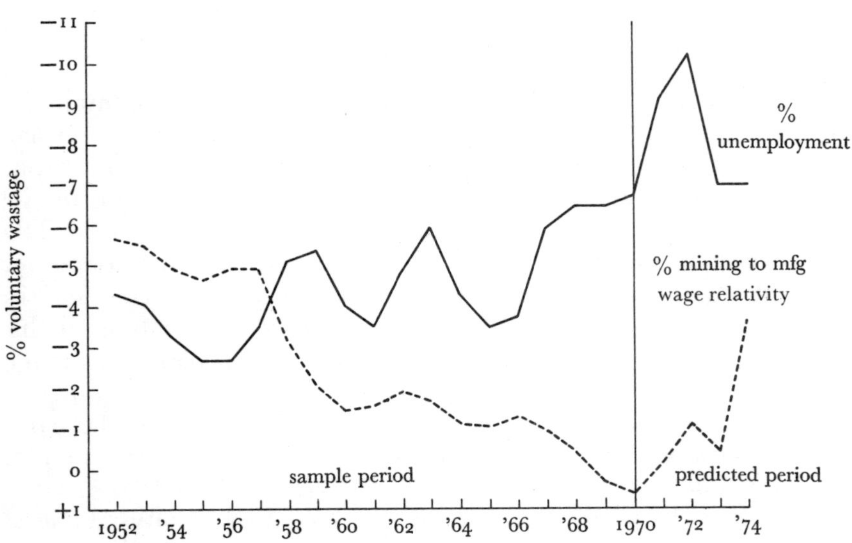

Figure 11.4 The contribution of unemployment and the mining/manufacturing earnings relativity to mining voluntary wastage 1952–74
Source: Department of Employment.

(5) should prove reasonably accurate as a predictive device. The voluntary wastage percentage predicted for 1974, on the basis of data from 1952–70, would have been 1 % lower than the actual percentage of 4.2. In numbers, one would have under-estimated voluntary wastage by some 2,600 men. The rather larger under prediction in 1973 is probably explained by the industrial unrest during that year, the construction boom, and a spate of pit disasters, all of which served to encourage men to leave the pits rather faster than might have been otherwise anticipated. Even so, the estimated voluntary wastage of 16,500 men is rather nearer the actual figure of 20,600 than the NCB's own estimate of 11,500.[15]

We also tested for a possible relationship between wage relativities and net changes in employment and recruitment. It is not surprising that, for the reasons we gave earlier, we could not detect a statistically significant relationship in either case. The fact remains, however, that in the longer term the trend of miners' relative wages moved in the same direction as the trend of mining employment – a relationship which is quite consistent with our findings. These results follow those of MacKay and his colleagues,[16] who found a strong positive relationship between voluntary wastage and the level of unemployment and relative wages in their engineering factories, but a much weaker relationship between these variables and net changes in employment or recruitment.

11.3 Output, manpower and pay relativities

Although we have shown one aspect of manpower supply in mining – voluntary labour wastage – to display some sensitivity to wage relativities with other industries, the question still remains whether, as the Pay Board argued in its report,[17] a substantial increase in the miners' relative wage was justifiable on manpower grounds. The answer clearly depends on the ability of the industry to obtain sufficient manpower – which depends in turn on recruitment in relation to wastage – in the light of its output target. It is to this question that we now turn our attention. The broader economic implications of our findings will be considered later.

Inland coal consumption fell from over 200 million in the early 1950s to around 130 million tons in 1973. Over the same period, deep-mined output fell by nearly 40 % and employment by over 60 % – the difference being accounted for by increased productivity. As late as 1969 coal was still considered to be a declining industry and the government was thinking in terms of a 90-million-ton deep-mined requirement in 1980.

The sharp increase in oil prices in the early 1970s suddenly changed coal's competitive position in the energy market. Instead of declining, the anticipation was that coal consumption would increase in the longer term. The concern now became to stabilise output in the short term and initiate an investment programme which would increase capacity to 150 million tons per annum.

In the accounting year 1972–73, NCB mines produced 127 million tons of deep-mined coal with a labour force of 268,000 wage earners.[18] At the beginning of the year 1973–74, the NCB[19] decided to stabilize output at around 130 million tons. It expected productivity at 472 tons per man year in 1972–73 to increase by 4% per annum. This would have implied a labour force of 265,000 at the beginning of the year, reducing to around 254,000 by the first quarter of 1974. Table 11.3 shows that employment was on target at the beginning of 1973–74 for a 130 million ton output. A sharp fall in the unemployment level in 1973, however, induced a voluntary wastage rate far higher than that anticipated by the NCB; some 21,000 miners left the industry for reasons other than deaths, retirements, dismissals and redundancy – 9,500 more than the NCB evidently anticipated. In addition, 15,000 miners left for involuntary reasons.[20] To maintain employment at the desired level, therefore, the NCB needed to attract approximately 26,000 recruits. Despite an intensive campaign only 15,000 men were recruited. Thus, by the end of the year, employment

Table 11.3. *Output and employment requirements 1972–76*

		Employment requirement at target output		Actual employment	
	Output per man year (tons)	120m. tons	130m. tons	beginning of year (000)	average for year
Assumed growth 4% p.a.					
1972–73	472[a]	254	275	274	268
1973–74	490	245	265	264	252
1974–75	510	235	254	243	245
1975–76	530	226	245	245	245
Assumed growth 2% p.a.					
1972–73	472	254	275	as above	
1973–74	481	249	270		
1974–75	491	244	264		
1975–76	501	240	259		

[a] *Digest of Energy Statistics* HMSO 1974.

had fallen to 243,000 wage earners or some 11,500 short of target. Even if productivity had risen by 4%, employment at this level would have been barely sufficient to obtain the revised target of 120 million tons of coal. In fact, productivity hardly increased in the months before the national strike. It seems doubtful, then, if the labour force at the end of 1973–74 would have been sufficient to maintain output at much more than 116 million tons. And if this employment trend had continued, the industry would have had a labour force of only 220,000 by the end of 1974–75. To have achieved 120 million tons of coal with a labour force of that size, productivity would have had to increase by 15% on 1972–73 levels – an increase unprecedented during the industry's post-war history.

One might argue that the industry's manpower shortage was only a temporary phenomenon, induced by the sharp fall in the unemployment level, although this was still higher than in any earlier recession. Our analysis would certainly suggest, for example, that the industry would have found it increasingly easier to retain and attract miners after the sharp rise in unemployment towards the end of 1975. To argue in this way, however, might suggest that people should be forced into coalmining by reducing alternative employment prospects. While there is certainly a case for not overheating an economy indefinitely it hardly seems a viable proposition to operate an economy at a level of unemployment which is compatible with securing the requisite number of miners; our evidence, for example, would suggest that if the miners' wage relativity in October 1973 had remained unchanged, unemployment would have had to be around 4.5%–5% to achieve that effect. Coal is a basic energy source upon which large sections of British industry are directly or indirectly dependent. In planning production targets one has to ensure that these will be sufficient to meet the energy requirements of a fully-employed economy. This being the case, wage relativities must be set at a level commensurate with obtaining sufficient labour to meet output requirements at a full capacity level of unemployment. While this necessarily implies maintaining a wage relativity above that necessary to ensure adequate supplies of labour when the economy is running below full capacity, this must be regarded as an unavoidable cost of ensuring sufficient capacity in the longer term. Our evidence indicates that, if we can regard full capacity as being commensurate with a general level of unemployment of about $2\frac{1}{2}-3\%$, miners' wages had reached a point, in 1973, where they were too low relative to those of workers in other industries.

We can elaborate on this point with the help of Figure 11.5, which shows the NCB's manpower surpluses and deficits between 1956 and

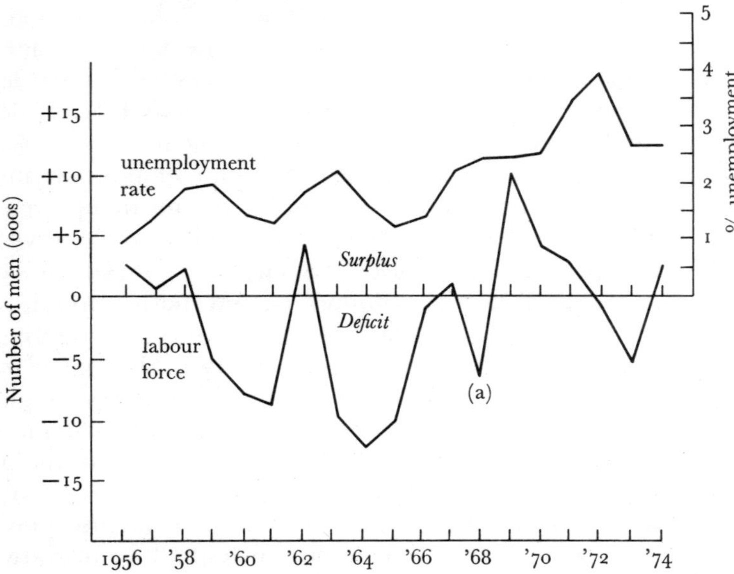

Figure 11.5 Mining labour force surpluses and deficits in relation to NCB manpower targets and the general level of unemployment
[a]The deficit in 1968 was produced by a rephasing of colliery closures.
Source: NCB.

1974. These are measured as the excess or shortfall of the average annual manpower on the industry's payroll in relation to the manpower requirement estimated by the NCB at the start of the year. This method of measurement means that an end-of-year labour deficit could be the result of one of two factors, or a combination of both, namely: an *inability* to recruit sufficient labour in the light of a given rate of wastage or, given labour availability, underestimation of the number of recruits required in the face of an unexpectedly high rate of wastage. Conversely, a labour surplus could occur because the NCB's intake of labour was based on an overestimate of realised wastage.

Figure 11.5 shows that for half of the years between 1956 and 1974 the NCB was operating with a manpower deficit despite the need to reduce the overall labour force. In part, the deficits, like the surpluses, undoubtedly reflect manpower forecasting errors and lags in recruitment and wastage. But the fact that, apart from 1968, when a deficit was produced by the rephasing of colliery closures, most of the deficits occurred in years of low (or falling) unemployment, and that in some years they were well in excess of 5,000 men, suggests that they also reflect a genuine inability to recruit sufficient workers to man the industry. Figure 11.5 also illustrates that, as the miners' wage relativity fell, labour deficits were occurring at increasingly higher levels of unemployment, even though, for most of the period before 1973, the industry had a declining manpower requirement.

While the sudden strengthening of coal's market position in the 1970s undoubtedly amplified the seriousness of the manpower situation facing the industry, one might argue that it only served to highlight many underlying problems which, even if coal had continued its earlier projected (declining) output path, would have increased the NUM's bargaining power and served to force an upward adjustment of the miners' wage relativity in the longer term.

First, there had been a considerable shift in the industry's regional distribution as output and employment were concentrated increasingly in the more profitable coalfields of Yorkshire and the East Midlands. In general, lower unemployment meant that the competition for labour from rival employers was more intense in these coalfields than in, say, Durham, Scotland, Wales and Lancashire. At the same time, the miners' external wage relativity in Yorkshire and the East Midlands had been eroded more substantially than in other coalfields. It is not surprising, therefore, that when manpower shortages appeared, they did so first in Yorkshire and the East Midlands. As coal output became even more dependent on these fields, it seems quite probable that it would have become increasingly necessary to raise relative wages. It is significant that in North Notts. and Doncaster – the two areas of most acute manpower shortage – employment fell below planned targets by 7.0% and 4.0% in 1973–74 as opposed to a shortfall of 2% in the other coalfields.[21]

The second aspect was the industry's increased reliance on craft skills as a result of mechanisation. The ratio of craftsmen to others employed, for example, had changed from 1:10 in 1960 to 1:5 in 1972–73. The skills acquired through the NCB's apprenticeship schemes can be utilised in a wide range of industries. As well as finding it increasingly difficult to attract young workers to undertake apprenticeships, the industry had difficulty in retaining qualified craftsmen, particularly in the inland coalfields of Yorkshire and the East Midlands.

The third aspect was the increasing age of the labour force. In 1958, 50% of miners were under 40, compared with 37% in 1971. This latter figure compares with 60% in construction – also a physically demanding industry – and 50% in manufacturing. It was partly the consequence of a manpower policy which had sought to run down the labour force by not replacing those leaving. But mining was also becoming increasingly unattractive as a career to school leavers and young adults. While the more secure future which the industry now offers might have helped to attract and retain younger men, this has to be offset against what appears to have been a growing aversion of younger men to manual work, of which mining is an extremely arduous example, and an increased reluctance on the part of mining families to encourage their sons to go into the industry. Both these trends have been encouraged by the growth of alternative employment opportunities in mining areas. At the same time, the elimination of traditional mining communities has reduced the potential manpower from that source. Thus the industry has been forced increasingly to compete in what we may term the 'open market' for manpower.

As well as being accepted by the Pay Board, many of the considerations outlined above in favour of improving the miners' relativity had been accepted in 1972 by the Committee of Enquiry established under the chairmanship of Lord Wilberforce.[22] The settlement negotiated with the new Labour government in March 1974 raised average earnings in mining by approximately 32%. This compared with an increase of 15% in manufacturing earnings over the previous year, and put average earnings in mining some 22% above those in manufacturing. In the short run, at least, this more than compensated for the deficiency left by the Wilberforce award; despite some differential erosion between March and October 1974, miners' average weekly earnings in the latter month were still some 6% above the relativity recommended by Wilberforce.

11.4 The impact of the 1974 settlement

The effect of considerably raising miners' pay relative to that of manufacturing males was, as we saw in Figures 11.1–11.4, immediate and dramatic – voluntary wastage slumped, thus easing the recruitment position. By the end of the accounting year 1974–75, voluntary wastage had been cut to nearly one half of its 1973–74 level, recruitment was above target, and the industry was able to stabilise its labour force at 246,000 men.

The extent to which voluntary wastage fell after the 1974 pay award did vary among major occupational groups. In no instance, however, was the reduction less than 40%. Table 11.4, which compares voluntary wastage rates among selected groups during the six months ending in September 1973 and September 1974, shows that, after juveniles, who have the highest labour turnover in all industries, the highest rate of voluntary wastage in the mines in the six months ending 29 September 1973 was among craftsmen, who clearly have the widest occupational choice of all miners, followed by faceworkers on the NPLA agreement, and other underground workers. The higher than average wastage among EBG workers is due to the high proportion recruited directly to these grades and thus to the higher rate of wastage among green labour. After the 1974 agreement, the largest percentage reduction in voluntary wastage was among NPLA faceworkers who received the largest absolute pay increase. The voluntary wastage rate among craftsmen also showed a substantial fall which, as in the case of faceworkers, was much greater than that of surface workers, among whom the reduction in voluntary wastage was least. Similar percentage pay increases were given to the under-officials (overmen, deputies, shotfirers) and weekly paid industrial staff as to faceworkers, and voluntary wastage among some of these groups also fell by an above-average percentage.

In addition to making it easier for the industry to achieve its overall physical manpower targets, the increase in the miners' pay relativity enabled the NCB to bias selection towards younger men, as well as to those with the better previous employment records (where the latter existed). In the case of new juvenile recruits, the NCB[23] found

Table 11.4. *Voluntary wastage percentage for occupational grades in 6 months ending 29 September 1973 and 6 months ending 28 September 1974*[a]

Selected occupations	Voluntary wastage (%) 6 months ending 29 Sept. 1973	Voluntary wastage (%) 6 months ending 28 Sept. 1974	Percentage change 1973–74
Overmen	1.25	0.53	–57.6
Deputy 1	1.71	0.64	–62.6
Deputy 2	1.80	0.94	–47.8
Shotfirers	2.14	1.26	–41.1
Other WPIS	1.01	0.47	–53.5
Fitters	4.75	2.42	–49.1
Electricians	6.33	3.23	–49.0
Other craftsmen	3.34	1.70	–49.1
NPLA	3.80	1.35	–64.5
Other underground (EBG)	4.98	2.78	–44.2
Surface	2.71	1.54	–43.2
Juveniles	14.94	7.74	–48.2
Others	9.54	3.72	–61.0
Total	4.37	2.08	–52.4

[a]Unemployment was fairly similar in these periods, so that the reduction in voluntary wastage can be ascribed to the wage relativity effect.

Source: National Coal Board, Recruitment and Wastage Returns, 1973 and 1974.

a fairly marked increase in the examination performance of applicants wishing to undertake craft apprenticeships over the previous two years although this may, in part, be the result of decreased job opportunities elsewhere. Thus, there is ample evidence to suggest that the enhanced external relativity enjoyed by the miners in 1974 had an immediate and predictable beneficial effect on the industry's labour supply, which could be measured in qualitative as well as in quantitative terms. Should the government wish to expand coal output and concentrate production in the more profitable English coalfields, which will include the as yet undeveloped Selby complex, therefore, it should be wary of a substantial erosion of the miners' external wage relativity, such as occurred during the early phases (1975–76) of the Labour government's incomes policy. The increase in unemployment between 1975 and 1976 served to depress voluntary wastage in coalmining, but shortages of workers in key craft grades were already appearing towards the end of the period.[24] This illustrates the problem of reconciling the interests of key sectors with those of the economy as a whole.

11.5 Summary and conclusions

Variations in the voluntary wastage rate in coalmining have been proved sensitive to, and largely explicable by, variations in the general unemployment percentage and in the miners' external wage relativity, measured as the percentage difference in average weekly earnings between adult males in mining and manufacturing. There was, however, no systematic relationship between either variations in the mining recruitment rate or net changes in mining employment in the short run and the level of unemployment or relative wages, whether the latter was measured as described above or as the difference between percentage changes in hourly earnings in coalmining and other industries. Numbers recruited (as distinct from the ease of recruitment) may seem insensitive to relative wages because recruitment levels are dependent on job openings, which depend in turn on the level of voluntary and involuntary wastage in relation to the industry's employment requirement at planned levels of output. Our findings do not end the debate on the role of relative wages in job choice and manpower allocation; coalmining might indeed be an exceptional case. However, they may suggest that comparisons of relative wage changes with net changes in the stock of employment, rather than with labour flows (namely recruitment and wastage of labour), are an inadequate guide to the sensitivity of labour mobility to relative wage adjustments.

What then has been the economic rationale of relative wage

adjustments in coalmining during the post-war period? The fall in the miners' external wage relativity towards the end of the 1950s coincided with a reduced demand for coal and manpower. So far as the NCB and the government wished to reduce mining employment with a minimal increase in unemployment, the fall in the miners' wage relativity – which encouraged voluntary wastage – was probably a useful adjunct of a manpower policy. However, our evidence also suggested that, as the miners' wage relativity fell after 1957, the industry found it difficult to attract enough recruits to replace involuntary and voluntary leavers at increasingly higher levels of unemployment, despite a falling manpower requirement overall. These trends were encouraged, no doubt, by the growing concentration of coal output in the more competitive labour markets and the industry's increased reliance on more general skills. One consequence of the wastage and recruitment pattern was a substantial ageing of the labour force. Without a fairly substantial wage adjustment, therefore, the industry's ability to meet its manpower requirements would have become increasingly dependent on the government's willingness to raise the general level of unemployment.

The employment problems which were clearly building up within the mining industry would probably have necessitated an upward adjustment in the industry's wage relativity in the longer term even if coal had continued on a declining output path. The sudden strengthening of coal's market position and the need to stabilise output, and possibly increase it in the longer term, however, gave added economic legitimacy to the miners' demand for a more immediate and substantial upward wage adjustment to halt the previously falling manpower trend. Whilst it is impossible to say just what relativity for miners would be justifiable on the grounds of manpower alone, it seems quite likely that the higher relativity established over the period 1974–75 (which corresponded to that prevailing in the late 1950s) would, if maintained, ensure sufficient miners in the medium term.

APPENDIX 1

	Average weekly earnings (October)			% Relativity	
	Coal	Manufacturing	Construction	Coal / Mfg	Coal / Con.
1948	8.47	7.7	6.53	118.1	129.7
49	9.08	7.40	6.85	122.7	132.6
1950	9.48	7.83	7.25	121.1	130.8
51	10.60	8.60	8.05	123.3	131.7
52	11.90	9.24	8.73	128.8	136.3
53	12.56	9.83	9.18	127.8	136.8
54	13.25	10.61	9.93	124.9	133.4
55	14.27	11.55	10.78	123.5	132.4
56	15.35	12.28	11.73	125.0	130.9
57	16.32	13.06	12.05	125.0	135.4
58	15.41	13.27	12.46	116.1	123.7
59	15.70	14.21	12.99	110.5	120.9
1960	16.28	15.16	13.95	107.4	116.7
61	17.16	15.89	15.25	108.0	112.5
62	17.93	16.34	16.10	109.7	111.4
63	18.75	17.29	16.63	108.4	112.7
64	19.73	18.67	18.20	105.7	108.4
65	21.21	20.16	19.77	105.2	107.2
66	22.16	20.76	20.56	106.6	107.8
67	22.92	21.89	21.68	104.7	105.7
68	24.12	23.62	22.87	102.1	105.5
69	25.10	25.54	24.46	98.3	102.6
1970	28.01	28.91	26.85	96.9	104.3
71	31.65	31.37	30.11	100.9	105.1
72	38.21	36.20	36.59	105.6	104.4
73	42.43	41.52	41.41	102.3	102.5
74	58.21	49.12	48.75	118.5	119.4

Source: D of E, *Gazette*

APPENDIX 2

	Average no. of wage earners on colliery books (000s) (1)	Voluntary wastage (000s) (2)	$\frac{(2)}{(1)} \times 100$ (3)	General unemployment rate (GB) (4)
1947	701.5	—	—	—
1948	713.2	40,874	5.73	1.5
1949	709.5	42,481	5.99	1.5
1950	687.4	50,798	7.39	1.5
1951	689.1	42,772	6.21	1.1
1952	706.2	31,963	4.53	1.6
1953	707.5	35,437	5.01	1.5
1954	701.8	37,384	5.33	1.2
1955	698.7	42,681	6.11	1.0
1956	697.4	39,630	5.68	1.0
1957	703.8	42,914	6.10	1.3
1958	692.7	32,872	5.32	1.9
1959	658.2	44,120	6.70	2.0
1960	602.1	57,953	9.63	1.5
1961	570.5	52,333	9.17	1.3
1962	550.9	33,552	6.09	1.8
1963	523.8	23,589	4.50	2.2
1964	497.8	35,013	7.03	1.6
1965	465.6	44,902	9.64	1.3
1966	426.8	39,893	9.35	1.4
1967	401.1	28,168	7.02	2.2
1968	349.5	29,175	8.35	2.4
1969	311.1	24,673	7.93	2.4
1970	290.3	22,398	7.72	2.5
1971	283.7	15,240	5.37	3.4
1972	271.0	9,673	3.57	3.8
1973	252.0	20,600	8.20	2.6
1974	246.0	10,400	4.20	2.6

Source: Ministry of Power etc., *Digest*, D of E, *Gazette*.

PART FOUR

SUMMARY, CONCLUSIONS AND POSTSCRIPT

12. SUMMARY AND CONCLUSIONS

We now summarise the main results of our study of wages policy and determination in the nationalised coalmining industry and consider their general implications. Apart from the historic importance, size and diversity of the industry, wage determination in coalmining since nationalisation is of special interest because of the attempt to revise and regulate a wage structure according to nationally agreed objectives against the background of rapid institutional, economic and technological change. Many students of wage determination and labour relations in the UK have emphasised the need to reform wage-fixing procedures and their associated wage structures if major problems in the field of pay and incomes are to be solved. The coalmining industry provides an opportunity to examine the causes and consequences of one such attempt.

12.1 Reform of the wages structure

The support which the miners' leaders gave to nationalisation of their industry was linked with their demands for the application of common principles in wage regulation. Before the Second World War, wage determination in coalmining had been wholly a local and district affair. The MFGB had managed to establish certain guidelines which the district unions might follow in coalfield agreements, but wages for the same job varied from pit to pit in a manner which defied all but the broadest generalisations. The coal owners had always insisted that national wage regulation was incompatible with their financial diversity. The long history of local autonomy in wage determination had been a divisive force in the industry and a major obstacle to the miners achieving a strong central trade union organisation which could negotiate with the coal owners on a common platform.

The miners' leaders expected that the transfer of the industry's assets to a single employer would be accompanied by a more coordinated approach to wage questions to facilitate overall planning. Machinery for national wage regulation had been established in 1942 and, in preparation for what it saw as an extension of national bargaining under nationalisation, the MFGB reorganised itself into the NUM in 1945. The scene was thus set for one of the most highly centralised collective bargaining frameworks in any British industry.

Since nationalisation, the NCB and NUM have moved the authority for wage regulation to the national bargaining table. Of particular importance in this respect was the series of wage reforms which established a single rate for every job in the industry. With an employer broadly sympathetic to the NUM's demand for more uniform treatment in wage matters, and obliged (by the terms of the 1946 Carry Over Agreement) to revise the industry's wage structure, it is perhaps surprising that it should have taken so long to achieve a revision which both the NUM and NCB thought would be accomplished not long after nationalisation.

In Chapters 2–5 of this study we outlined the evolution of a national wage structure in the industry, this being of fundamental importance to our subsequent evaluation of wages policy. Our purpose was to show that, whereas in the longer term the NCB and NUM have sought to revise the wage structure in a direction compatible with the egalitarian objectives of the NUM, they have done so subject to economic, technological and institutional constraints and considerable outside interference.

Despite a commitment upon nationalisation to undertake an early reform of the industry's wage structure, neither the NCB nor the NUM had a detailed blueprint for revision. The NUM was committed, in principle, to achieving greater equality between men doing similar tasks. But it was also aware of the opposition which such a policy might encounter from the higher-paid areas if it was not pursued in a manner which seemed fair to all concerned. This put a budgetary constraint on the union's aspirations and the NUM was wary lest the cost involved in achieving greater wage equality should jeopardise advances on the more general wages front and, in particular, its wish to improve the relative position of the lowest paid. Also the union was pledged to end piecework. But the pressing need for coal, and the dependence of output on human endeavour, militated against this objective in the short term. The NCB was concerned that too radical an alteration of the internal wage structure in a situation of manpower shortage might, and probably would, jeopardise coal supplies. Thus reform of the industry's wage structure had to be set against other, often conflicting, priorities and considerations, which, taken overall, favoured a gradual and somewhat pragmatic approach to wage reform, which ultimately followed the lines of least resistance.

12.1.1 *The daywage reform*

The first major revision of the wage structure, outlined in Chapter 2, was implemented in 1955, after the National Reference Tribunal placed considerable pressure on the NCB and NUM to devise a detailed blueprint for reform. Although the wage rates negotiated in 1955 were applicable immediately only to daywagemen, the national grading scheme was acclaimed as the first step in ensuring that every man in the industry would be paid for the job he performed, irrespective of the geographical location of his pit.

The wage-rate structure established by the 1955 agreement was influenced partly by the existing pattern of wage rates, and partly by the financial resources available for revision. The NCB would have preferred larger differentials between the grades to reward the acquisition of skills, but this was impossible, given the financial constraint. The agreement was the result of a systematic survey of existing wage rates and job definitions. Care was taken to ensure that no man suffered a wage cut so long as he remained in the same pit doing the same job, and the agreement provided that most daywagemen would get an increase. In this way, majority support within the NUM was ensured, even though the agreement was opposed in some of the coalfields where relatively highly-paid miners stood to gain least in the medium term.

12.1.2 *The piecerate reform*

The 1955 agreement left the piecerate structure virtually unchanged. It was next in line for revision, but the system, details of which were described in Chapter 3, revealed more problems than had been envisaged.

First, because a nationally consistent piecerate structure was unlikely to raise wages faster in the lower-paying areas, it would have been unacceptable in the relatively militant and low-paying areas, such as South Wales and Scotland. Second, earnings norms were being agreed unofficially at district level in an attempt to reduce the spread of rates in particular coalfields. This distorted the effort–wage relationship still further, but it offered greater scope for controlling aggregate wage increases. Third, the cost of revising the system became increasingly prohibitive as the industry entered a period of sharp competition from cheap imported oil. Fourth, the replacement of conventional mining by power-loading machines provided an opportunity to negotiate new wage agreements for faceworkers. As power loading progressed, these agreements gradually replaced the old piecerate system.

Discussion on the reform of the 'old' piecerate system continued in the late 1950s and early 1960s, but attention shifted to the possibility of negotiating a national agreement for power-loading operators. In Chapter 4, we explained how these early negotiations foundered on the NUM's refusal to accept that task norms – and thus piece rates – should be based on method study. Meanwhile, district officials of the NUM and NCB were left to negotiate their own power-loading agreements, of which all except one were concluded on a district basis and tailored to meet the circumstances of particular coalfields.

When discussions on a National Power Loading Agreement were re-opened in 1965, consideration was at first given to the possibility of achieving a national *piecework* agreement, both sides being seemingly convinced that piece rates were still applicable to power-loading operations. A national daywage agreement had been concluded for operators of the more sophisticated ROLF installations in 1965, however, and after discussions with officials of the Donovan Commission and the National Board for Prices and Incomes, who were concerned about the association between PBR systems and wage drift and unofficial strikes in industry more generally, the NCB changed to favouring a daywage agreement for power-loading operators. Thus, just as the first major wage reform of 1955 owed much to the pressure which the National Reference Tribunal had applied to the NCB and NUM, so other independent bodies were instrumental in fashioning the 1966 National Power Loading Agreement, which marked the second major step in national wage structure rationalisation.

The NPLA was drawn up in great haste. This had the advantage of surprise, which probably helped officials of the NUM, such as Will Paynter (the General Secretary), who supported the move to a uniform daywage system, to obtain a small majority in favour of acceptance (however, this was mainly ensured by fixing wage rates at levels which were attractive to sufficient workers in the short term). The NPLA did not establish uniform wage rates for power-loading teams overnight. Parity was to be achieved over a period of five years, during which face teams in the lower-paying areas were to receive considerably smaller increases than those in the higher-paying areas.

The NPLA set the precedent for the complete abolition of piecerates in coalmining. An agreement similar to the NPLA was concluded for the remaining pieceworkers and taskworkers performing jobs behind the coalface in 1971, by which time uniform standard wage rates had been fixed for the NPLA operators throughout the industry.

Our account of wage structure illustrated that positions on the nature of wage reform shifted quite considerably in the short run, according to economic and technological developments and prevailing beliefs about the efficacy of various payment systems. Thus, while in the longer term wage structure reform in the coalmining industry apparently reflects the gradual imposition of institutional and egalitarian ideals over contrary economic arguments, it has not been a mechanistic process, with one or both parties dogmatically pursuing immutable policies. Negotiations have involved both parties trading off various objectives and arriving at solutions which were politically and economically acceptable at the time.

12.2 Central wage regulation

The wage reforms outlined in the first section of this study were one aspect of a more general attempt by the NCB and NUM to regulate the mining industry's wage system according to nationally agreed objectives. Just how effective they have been in this respect has been the subject of considerable controversy. We noted in Chapter 6, for example, that several commentators have identified a persistent wage drift, until the mid 1960s at least, and doubted the impact of policy in regulating wage differentials between the highest- and lowest-paid groups. These arguments were also used by the NCB in its justification of the decision to abolish piecework in 1966. Our analysis has cast some doubt on the validity of the assumptions upon which these assessments have been based.

12.2.1 *Measuring the wage drift*

Previous attempts to measure the wage drift in mining take virtually no account of the intentions of the national bargainers in different phases of wages policy and have paid scant attention to the data on which the measures have been based. It is inevitable, therefore, that they should have given a misleading impression of the efficacy of national decisions in scheduling actual wage movements. Most students of coalmining wage movements, for example, have estimated the drift as the difference between percentage increases in what is effectively the minimum wage rate and actual EMS. This calculation assumes that actual EMS *should* have risen by the same percentage as minimum wage rates. This could approximate to an observed tendency but, as we have shown, it reflects neither the intentions of the central bargainers nor the actual mechanics of wage determination over most of the period considered.

Prior to the abolition of piecework, the coalmining industry had formal provision by *national agreement* for local and district bargaining. The purpose of national agreements was to lay down rules governing pay increases at these levels. They could not hope to determine actual pay increases, because they could not determine all the factors, such as attendance, productivity and working conditions, which could influence their magnitude. Furthermore, by ignoring such nationally agreed items as attendance bonuses, holiday and sickness pay, allowances for working in abnormal conditions, and overtime pay – all of which are included in the published statistics of miners' pay – previous writers have seriously underestimated the contribution of national decisions to actual wage movements, particularly in recent years when the relative importance of many of these 'supplementary payments' has increased quite substantially. However we measured the impact of national agreements on pay movements, our calculations in Chapter 6 proved that national decisions had become increasingly effective in regulating wage movements.

12.2.2 *Piecework bargaining*

Several commentators argued that piecework was a destabilising force in the coalmining industry's wage structure and primarily responsible for the observed wage drift before the late 1950s. The same writers suggested that the percentage shift rate differential between pieceworkers and daywagemen widened over the same period. Our analysis in Chapters 6 and 7 did not wholly support these contentions.

Local piecework bargaining – mainly on allowance payments – undoubtedly posed a problem for central wage control in the absence of adequate wage-vetting machinery and given the manpower deficiency during the early years of nationalisation. By 1956, though, the percentage wage gap between pieceworkers and daywagemen had declined to a level below that of 1948.

The percentage wage gap between the two groups did widen slightly in the late 1950s, but thereafter central wages policy seems to have been increasingly effective in regulating pieceworkers' EMS, which increased *less* fast than that of daywagemen in percentage terms.

Our analysis suggested that national wage increases to daywagemen helped to determine the upper limit on percentage increases to pieceworkers from local bargaining. However, it also suggested that, with the exception of the late 1950s, the differential between the two groups contracted in percentage terms because of increases in overtime pay and a small element of wage drift in daywage earnings. In short, whereas pieceworkers were able to match nationally agreed increases to daywagemen which they could estimate quite easily, they could not wholly predict, and offset, increases in the variable element in daywage earnings. This would suggest that local piecework bargaining was to some extent constrained indirectly through national wage decisions, although the latter were only partly responsible for the differential contraction which took place between pieceworkers and daywagemen.

If wages had moved in line with the egalitarian intentions underlying national agreements, EMS should have risen fastest in the lowest-paying coalfields. We found in Chapter 7 that there was some evidence of this before the NPLA, but that this trend was confined mainly to the daywage structure. The pieceworkers' structure did contract slightly during the early 1950s, but in the mid 1950s it actually widened, although this was mainly the result of wage movements among face pieceworkers rather than EMS movements of pieceworkers elsewhere underground, which tended to follow the pattern set by the EMS changes of daywagemen. There was some evidence of a widening of the percentage differential between face pieceworkers and surface daywagemen in many of the higher-paying coalfields in the late 1950s, but this was due, in part, to the effective application of the ceilings policy on daywage earnings in those areas before 1958.

Our analysis in Chapter 7 suggested that variations in the rate of increase in pieceworkers' EMS, which caused coalfield EMS differentials among face pieceworkers to widen in the late 1950s, were associated with the uneven introduction of power loading in the coalfields, the decision to pay power-loading operators higher rates than faceworkers operating conventional machine faces, the larger growth of face productivity which accompanied power loading, and differences in the style of agreement adopted. By the early 1960s, however, there was some reduction in percentage wage differentials between face pieceworkers; the differential between face pieceworkers and surface daywagemen was reduced in every coalfield except Nottinghamshire before the NPLA.

In view of the strong upsurge in productivity following the spread of power loading, it is perhaps surprising that pieceworkers' EMS did not rise faster and that it was possible to effect a reduction in the pieceworker/daywage differential in such circumstances. The ability to do this was helped, no doubt, by the declining manpower requirement and the competitive economic environment in which coal was operating. Increased cost consciousness after the mid 1950s may also help to explain the closer association between face pieceworkers' EMS increases and productivity changes between 1956 and 1966 than in the previous decade.

In only two coalfields did EMS rise faster than productivity between 1956 and 1966. If, as seems to have been the intention of national wages policy, we judge the efficacy of local wage control procedures by their ability to relate wage increases to increases in productivity, the object being to stabilise wage costs, our evidence suggests that wage regulation on piecework contracts was more effective than many commentators, simply looking for contracting tendencies in the piecework wage structure, have so far admitted.

12.2.3 *The daywage structure*

We noted several difficulties in administering the daywage structure after 1955. There was undoubtedly fairly strong resistance in the higher-paying areas to the spirit of the agreement. In Chapter 5, we illustrated how this resistance was accommodated in part in national negotiations by a raising of wage ceilings so as to extend the period in which personal-rated men would be brought on to standard wage rates. But even after this concession, there were other problems in administering the daywage structure after the late 1950s. Some of these may be attributed to the failure of the national negotiators to adjust the structure to the changing needs of the industry.

Although the 1955 grading structure provided a framework in which national wage policy decisions could be systematically applied, no one expected it to remain unchanged; new jobs following technological innovations could have been added and the wage structure adapted to reflect changing skill content (and presumably manpower priorities). The wage-structure planners had expected that larger differentials would be established between the various grades. But the NCB yielded to NUM pressure to raise the wage rates of each grade by uniform money additions. This demand was governed partly by equity considerations and also by the hope that the notion of flat-rate increases would spin off into piecerate negotiations, thus narrowing the percentage wage gap between pieceworkers and daywagemen. While our analysis suggested that the policy had some effect in this latter respect, it inevitably eroded wage differentials between the daywage grades, which ultimately became meaningless. The daywage structure also became less differentiated as the grading structure was rolled-up in an attempt to help the lowest paid.

Craftsmen were the one exception to the general policy outlined above. With more advanced techniques, the industry became increasingly dependent on engineering skills which were in short supply, relative to demand. In the late 1950s, the industry experienced difficulties in recruiting and retraining craftsmen which, as the craftsmen became an increasingly important group in the NUM, led the union to agitate for an increase in the craftsmen's differential over other daywagemen. The NCB could not afford to include all craftsmen in this differential adjustment and chose, instead, to be more selective in favour of fitters and electricians, of whom there was a particular shortage in the inland coalfields. This created anomalies in the craft grading structure which, together with those resulting from the NPLA, resulted in a wholesale revision of the craftsmen's wage structure in 1971.

Technological changes also affected the jobs of some non-craft daywagemen. But these changes were catered for only sporadically and the Black Book became increasingly out of date, leaving local management scope to adjust to local pressures. Overtime, task rates and upgrading all appear to have been lowest in the less viable coalfields. Thus local deviations from nationally agreed rates were positively related to differences in the economic viability of the coalfields (which became greater as the industry faced a falling demand for its product) and inversely related to national wage-rate increases. These problems were also manifest in the problem of controlling overtime at the coalface after the NPLA. Thus, while national wages policy was trying to push the wage structure in one direction, there were local pressures tending to push it in the opposite direction.

12.3 External comparisons

Comparisons between wage movements in mining and other industries are difficult to make, because of differences in data reporting, which we described in detail in Chapter 8. Our results on external relativities were rather different from those of previous commentators, who claimed, without distinguishing between face and surface workers, that, after a sharp fall in their external relativity in the late 1950s, miners maintained their reduced relativity in the early 1960s.

During the first decade of nationalisation, the sharp contraction in the daywage structure occurred against the background of what, helped by the small element of wage drift and increased overtime working, appears to have been a slightly improving external advantage for daywagemen as a whole. Pieceworkers maintained, but did not improve upon, their external wage position before 1956. The external relativity of all miners fell in the late 1950s; but thereafter the daywagemen managed to maintain their reduced relativity in hourly terms, while that of pieceworkers and face pieceworkers, in particular, continued to decline until 1971.

Several factors explain why, even before 1966, pieceworkers were persuaded to absorb the major burden of coal's declining market fortunes. First, they were relatively highly paid. Second, the alternative jobs available to them were relatively low paid. Third, management, particularly in the less viable coalfields, could afford to resist their demands for high wages so long as there was a declining demand for their services. We might thus explain the decline in the faceworkers' internal and external relativity in terms of their weakened bargaining power, which was reflected in their declining strike frequency. Thus, economic trends tended to endorse the internal wage objectives of the national negotiators.

The same factors help to explain why, despite further substantial erosion of both their internal and external differential, faceworkers appeared to be acquiescent during the early years of the NPLA. But such an explanation is probably too simple. What was happening during the early years of the NPLA, and before it for that matter, was that some faceworkers were faring better than others. This tended to camouflage the relatively unfavourable underlying trend of earnings. By the terms of the NPLA, the NCB was offering faceworkers security of wages, in return for what, in many cases, was a substantially reduced standard of living in the medium term. As the number of workers operating under the agreement increased, their opposition to it became more unified, and this was made worse by keeping the highest rate of wage increase well below the increase to daywagemen, in the interests of reducing wage differentials between face and surface grades.

The substantial erosion of the faceworkers' differentials in the later 1960s, and to a smaller extent that of daywagemen in the high-paying coalfields, was an important factor in the miners' successful claims for an increase in their external relativity in the early 1970s. But while the NUM were asking for an increase in the external relativity, they were seeking to contract the internal wage structure still further. Thus, encouraged by coal's strengthened market position following the sharp increase in oil prices in the 1970s, the NUM were using the results of their egalitarian policy as a lever to raise the whole wage structure. It is not surprising, in view of the probable repercussions of such a policy on the rest of the economy, that the NUM should subsequently have been under considerable pressure from the NCB to reappraise its wages policy in favour of a widening of differentials between faceworkers and other miners.

12.4 Productivity

We examined many of the arguments used to justify the NPLA in 1966 and found most of them to be dubious. In particular, we suggested, technology was not at a stage in 1966 where output could be said to be independent of human effort; nor were face pieceworkers' wages generally rising faster than those of the lower-paid daywagemen.

There is an implicit assumption in any move towards wage uniformity that the requirements of the job and the ability of the worker performing it are synonymous. Common sense tends to contradict such an assumption. One thing which did emerge from our analysis of the old piecerate system on handgot faces, for example, was that, although piece rates varied for reasons which did not always correspond to effort, there was nevertheless some general tendency for them to be related. This was an obvious embarrassment in any discussion of uniform piece rates, for it meant that actual earnings could have fallen in just those areas where the pressure was to raise them. Our evidence on the effort/wage relationship on power-loading faces was less detailed, but it is unlikely that task standards were sufficiently variable to account for the wide differences which existed in EMS in the piecerate coalfields. If this was so, men might have felt justified in reducing their work rate where their earnings suffered in the process.

This was not an easy proposition to test. The changeover from piecework to fixed daywages in coalmining did seem to be associated with some reduction in effort input. There was also evidence that, at coalfield level at least, differences in productivity changes after the NPLA were inversely associated with differences in the rate of faceworkers' wage increases. This suggests that the realignment of wage rates following the NPLA provoked some realignment of effort input. In some coalfields, therefore, 'wage drift' was replaced by what we might call an 'effort drift'. There was also a fairly sharp increase in overtime working in the coalfields where wages were to rise least in the six years of the NPLA phasing in of wage rate equality. Thus, faceworkers in the previously higher-paying coalfields, perhaps in collusion with management, were able to obtain some compensation for their declining inter-coalfield relativity by working more hours at overtime premium rates. However, the growth of overtime was exacerbated by the problem of recruiting faceworkers in many such coalfields as their wage relativity declined.

Although the changeover from wage incentives to fixed daywages was not wholly responsible for stagnating productivity in the early 1970s – the lower rate of technological change, adjustment to an increased demand and less rapid closure of the least productive pits were all important factors – the attempt to raise productivity to meet a probable increase in coal demand encouraged the NCB to contemplate reintroducing some form of wage incentives scheme in the industry. The problem now is to reconcile the wish of an important section of the NUM's leadership to retain national authority over pay and use the faceworkers bargaining strength to raise the whole wage structure, with the need to provide some local incentive to raise productivity.

12.5 Strikes and wage reform

The sharp decline in the mining industry's strike frequency in the late 1950s has frequently been associated with the spread of power loading and the new wage agreements which governed pay on those machines. This in turn has been used to support a more general application of pay systems which give more stable earnings. While there was some evidence to suggest that the task-rate agreements applicable in Scotland and parts of Yorkshire encouraged the reduction in the frequency of strikes in those coalfields, the reduction in the industry as a whole, which can be clearly pinpointed to 1957, seems mainly a reaction to the fall in coal demand which occurred in that and subsequent years. This view is supported by the fact that the incidence of issues going through the conciliation procedure actually increased in the late 1950s.

The NPLA was certainly instrumental in reducing the frequency and proportion of strikes over wage issues after 1966. But this favourable effect was partly offset by disputes over working arrangements, such as manning and job flexibility, which the agreement was intended to encourage. This supported our findings that bargaining at local level switched from the wage *per se* to the effort input at a given wage. Complaints that the agreement had reduced the authority of local lodge committees were also quite common.

The main effect of the NPLA on strikes was to unite the NUM. The NPLA caused dissatisfaction among miners in the traditionally higher-paid coalfields. Direct production workers could now see more clearly their dependence on the national negotiators, so were more willing to give their support to a common platform. The 1970s saw the first national strikes in mining since 1926. More working days were lost in the two national strikes of 1972 and 1974 than by the sum total of local (unofficial) strikes in the whole period of nationalisation. While it was part of a more general phenomenon, the growth of national militancy in mining was also associated with the introduction of a revised wage structure for pieceworkers, and the deleterious effect this had on their standard of living. From the standpoint of industrial relations, therefore, the mining case is a stark reminder that the efficacy of procedural reforms of collective bargaining cannot be evaluated independently of their substantive consequences.

12.6 Wages and labour mobility

There has been considerable debate among economists about the effect of wage relativities on labour mobility. In Chapter 11 we were able to point to several fallacies in earlier attempts to test the sensitivity of labour mobility to relative wage changes from data which relate percentage changes in net employment to average hourly earnings. Net employment changes take no account of the flows of labour into and out of an employment. It is the magnitude of these changes, and their composition, which are important for both wage theory and employers' attitudes to wage relativities. Of these flows voluntary wastage is the variable most likely to be sensitive to economic factors – both involuntary wastage and recruitment depend on factors which are not wholly dependent on the individual worker's discretion.

Voluntary wastage from coalmining over the post-war period was shown to be affected by relative wage changes. The percentage difference between average weekly earnings of coalmining and manufacturing adult males was found to be the most relevant variable when measuring wage relativities. But voluntary wastage was found to be more sensitive still to changes in the general level of unemployment, this being a crude measure of available job opportunities. As it seems reasonable to assume that workers seldom move *voluntarily* to lower-paid occupations, however, this should not be taken as proof of the job opportunities hypothesis of labour mobility over the wage relativities hypothesis. Although we could find no positive association between external relativities and recruitment to coalmining, it again seems fair to assume from our voluntary wastage equation that the ability to recruit is similarly sensitive to both relative wages and the level of unemployment.

Our findings provided a useful insight into the rationale of the NUM's claim for a substantial increase in the miners' external relativity in the late 1960s and early 1970s. So long as the demand for coal was falling and productivity increasing, the NCB could afford to let the miners' external relativity decline, although this effect was produced in part by the sharp cutback in available overtime hours in the late 1950s and early 1960s. This encouraged voluntary wastage from the industry which, with a restriction on recruitment, helped the industry reduce its labour force to levels which were compatible with a reduced demand for coal. But as the younger men tend to be the most mobile, this produced a marked ageing of the mining labour force. Because of the industry's increased dependence on craft skills, and the wider labour market for those skills than for other categories of mining labour, however, the NCB was forced to give more favourable consideration to national wage claims on behalf of craftsmen than on behalf of other workers.

By the late 1960s and 1970s, though, shortages of miners in non-craft occupations were beginning to appear as soon as the economy approached full employment output levels. The most serious shortage was of men to operate power-loading machines. This was being aggravated by the sharp contraction of the internal wage structure, which reduced the attraction of moving to the coalface. Further, it was most acute in the more competitive labour markets of the East Midlands and Yorkshire, where coal production was being increasingly concentrated. Thus, the NUM's hand in wage negotiations was being strengthened independently of the sharp reversal in coal's market position following the sudden increase in oil prices.

The need to stabilise coal's output potential in the 1970s made the case for a substantial increase in the miners' external relativity difficult to deny. Despite the threat which such an adjustment would pose to the government's attempt to operate an incomes policy, it could be argued that it was a pity that the government did not recognise the strength of the miners' position early enough either in the disputes of 1971/72 or 1973/74. If it had done so, lower levels of settlement would almost certainly have been possible. However, the economic policies pursued by the government had brought about, by the summer of 1973, a tightening of the labour market which was bound to create enormous pressures on an inflexible incomes policy. The construction boom of the early 1970s and the wage escalation in that industry, which was a special feature of that boom, had an especially harmful effect on the coalmining industry's ability to retain its labour force and attract new recruits, at a point when it needed, for the first time in many years, to stabilise its manpower levels. The success of central incomes policy is dependent in part on the government avoiding excess demand and sharp oscillations in economic activity. It also follows that there are dangers inherent in the wages of a key sector moving out of line with those of other sectors of the economy, even when this is the inevitable result of internal wages policy. This again emphasises the desirability of a gradual approach to changes in relative wage levels, and avoiding the need for sudden re-adjustments. It may also help to explain why employers often seem reluctant to change wage structures. The problem for the coal industry now, and indirectly for the government, is to achieve a wage structure for coalmining which is consistent with the production demands which are likely to be made on the industry in the next decade, without provoking wage inflation in the rest of the economy. This will require a change in the NUM's internal wages policy.

12.7 General conclusions

It is dangerous to generalise from a study of wage behaviour in one industry, especially in view of the current desperate search for general panaceas in wage determination. This statement is intended to go beyond a formal cautionary note. Many aspects of wage behaviour in coalmining can only be understood in terms of influences which are specific to the industry. It is highly unlikely that, without a relatively strong trade union with a fair proportion of egalitarian-minded leaders and the transference of the industry's assets to a single public employer, able to adopt an overall approach to production and manpower planning and cross subsidise its operations, coalmining would have evolved a common wage scale for its labour force. National wage structure developments in coalmining thus owe much to the institutional framework of the industry.

An attempt to reorganise a pay system and structure is usually a step into the unknown. An existing system may not be the most efficient, but the repercussions and cost of revision can be unexpectedly high in the short run and change may jeopardise the supply of key labour in the longer term. Employers and unions may have 'some feel' for the labour market, but they can only test that appreciation against the consequences of their actions. Also, in the coal industry, and this may apply generally, they respond to short-term pressures, and perhaps fashions, without fully appreciating, or necessarily being able to predict, the long-run consequences.

The assumption that uniform wage rates could be imposed and maintained in the coal industry, which is, despite common ownership, still diverse and geographically dispersed, certainly contradicts traditional wage theory with its heavy emphasis on supply and demand, since labour market conditions have always varied considerably between coalfields and the pits within them. In moving towards an egalitarian structure, the central negotiators tried to accommodate perceived labour market pressures. For example, although some men suffered a wage cut as a result of revision, the emphasis was on the levelling up of wages. When pressures emerged as a result of the levelling up process, wage ceilings were raised to accommodate these pressures. When shortages of craftsmen emerged, wage-rate increases were differentiated insofar as institutional pressures would allow, to take account of them. These were all devices to absorb labour market pressures within a uniform framework. This experience indicates the difficulties likely in any attempt to impose a standardised system of wage rates in other industries, many of which are more diverse.

Despite the attempts to accommodate perceived labour market pressures, it would be idle to pretend that the coalmining industry has a perfect record on the question of wage regulation. Central policy has undoubtedly been most effective in the less profitable, and thus more cost conscious, coalfields. This probably reflected an allegiance to tradition as much as to genuine forces of supply and demand in the labour market, but there can be no doubt that it was helped by the differing implications of wage equalisation for employment in the areas of relatively high and low profitability. Thus, central wages regulation did meet a varying response in the coalfields, which served partly to offset the equalising objectives of policy, further illustrating the problems of imposing uniformity amidst diversity.

The efficacy of wage regulation procedures and the workings of the labour market cannot be judged by mere short-term adherence to national wage agreements. And it could be that labour economists have given far too much attention to the wages aspect when studying how the labour market works. The mining case illustrates that a worker has several options open to him, other than negotiating for more money. He can leave his job, he can work at a slower pace, he can switch his hours of work between overtime and normal hours, and change his leisure/work pattern. These are all elements of behaviour which affect the supply of labour and may *ultimately* affect the wage rate. Even if institutions are able to regulate a wage structure in an administrative sense, it is unlikely in our society that the behaviour of the people who work for that rate can be completely controlled. The actions of individuals may thus distort a collectively-agreed wage structure when it is at odds with labour market conditions. The institutional response in the longer term must be to accommodate those pressures which conflict with the economic priorities of an organisation, or institutions will find themselves overtaken by them. Wage behaviour in the coal industry has been influenced very substantially by collective actions, but it also reflects the behaviour of individual workers (and managers).

It is most unlikely that a wage structure is ever in equilibrium at any moment in time: the consequences of individual reactions to sets of economic forces may take time to reveal their full extent. For practical purposes this probably means that there is a margin of flexibility, within which institutionally imposed changes of structure can occur without undesired side effects. But no organisation can afford to operate completely independently of others. Nor can it regulate every aspect of individual behaviour in the way which would be necessary for it to achieve its desired goals without generating offsetting repercussions. What applies to the coal industry applies even more to the economy as a whole.

Thus, while institutions can have an impact on internal wage structures, there are clearly defined limits within which they have to operate and within which their policies can be made effective. These limits are imposed by the interests vested in the existing wage structure, the product market of an industry, technological developments, wage movements in the rest of the economy and, best seen as reactions to each of the variables, the behaviour of individual workers and managers whose perceptions of how the labour market works may be very different from those of central wage administrators.

The need to reform collective bargaining procedures and associated wage structures has been a continuing theme in the literature on industrial relations and wages policy. The picture of wage systems riddled with anomalies, prone to industrial unrest and leapfrogging pay claims, over which the parties to collective bargaining have little or no effective control, is now all too common. Only (so the argument goes) when we have adequate procedures for the negotiation of agreed rules and regulations to govern pay and related conditions of employment, are we likely to see any lasting solution to many of the problems in the field of pay and incomes.

This view was forcibly expressed by the Donovan Commission, in its report[1] of 1968 on industrial relations in Great Britain. The Donovan Commission also took the view that the deficiencies which it described at some length could only be dealt with at the level of the workplace or company. It saw the basic defect of the present system to be the inability of industry-wide collective agreements to accommodate pressures in the place of work. Those pressures were thus absorbed in an ad hoc manner and it was from that 'ad hocary' that most of our problems derived. The solution offered by Donovan was to shift the emphasis for collective bargaining to the level of the company and plant and to provide and improve procedures at those levels.

No-one has ever been quite sure where the nationalised industries, like coalmining, fitted into the Donovan analysis, and recommendations for them were virtually ignored in a report which dealt mainly with the private sector. Donovan's argument for their exclusion might simply be that the agreements of (say) coalmining are really company agreements and thus fit into the favoured framework. But if this is so, the report would seem to ignore many of the problems which confront employers and trade unions in a company framework. It would also seem to be based on a misunderstanding of what national agreements seek and can hope to achieve.

This study suggests that the problem of wage determination, and the appropriateness of various institutional bargaining frameworks for it, is by no means as simple as Donovan claimed. The case for decentralised, as opposed to centralised, collective bargaining hinges on the need for wage flexibility to accommodate local labour market pressures. It remains a fact that, whatever the bargaining framework, wage regulation has to be combined with sufficient flexibility to meet labour market pressures in an orderly fashion. In a highly centralised and uniform wage structure, this may mean that wages have to be revised above the equilibrium rate in some areas, to accommodate pressures in those where there is a problem of labour supply. This can obviously be a costly structure.

But a trade union like the NUM may not accept a decentralised bargaining framework. Thus institutional reform of collective bargaining cannot be divorced from the substantive objectives of trade unions and employers. The miners and their unions have surely displayed an infinite capacity for adapting their demands to the changing circumstances of their industry. They have been able to trade off various objectives in both the short and long run, as best suited them in the institutional, economic and technological climate in which they were forced to work over the period considered. We have little reason to suppose that other work groups are incapable of doing the same. The mining trade unions, like unions in other industries, see the system of collective bargaining as existing for a purpose. When it failed to provide the particular result they had in mind, it was circumvented, even to the point of direct confrontation with the government. No system of collective bargaining is immutable, nor is any wage structure. If the miners agree to return to some form of local bargaining, albeit within a national framework, it will be because it gives them more chance of raising pay in the prevailing economic climate than would otherwise have been the case. But this may only serve to emphasise that organisations tend to evolve bargaining frameworks which best suit their particular purposes, which will depend in turn on the environment in which they are operating. A reorganisation of collective bargaining procedures *per se* does not solve many of the basic problems associated with determining pay. Within any bargaining framework, one still has to make a decision about the pace at which wages should increase and how that increase should be allocated between occupational groups. It is these questions which seem to give rise to most of our problems over pay and to which we have directed our attention in this study. Without attempting to generalise too much from the mining experience, one is forced to conclude, therefore, that much recent conventional wisdom on collective bargaining – especially as to the desirability of certain forms as against others – was both inadequately informed and altogether too sanguine in its expectation that mere institutional changes could solve major problems of pay and incomes.

13. POSTSCRIPT: WAGE INCENTIVES 1977–79

The account of wages policy in the coalmining industry in the preceding chapters was completed during the summer of 1977, although due to the time lag of data publication many of the statistical series conclude at a somewhat earlier date.

There is always a danger with a subject of topical interest that events between the date of writing and publication will prove the writer wrong and thus necessitate a major reappraisal (and rewriting) of the work. I consider myself fortunate and am reassured by the fact that subsequent developments on the coalmining industry's wages front have conformed to my predictions. In this context I refer especially to the miners' gradual acceptance of productivity-related wage incentives after November 1977.

Rather than up-date the relevant statistical tables of each chapter in the light of the impact which the introduction of a productivity payment scheme has had on such things as general wage movements, internal wage differentials, external relativities, productivity, output, disputes and accidents, it seemed more appropriate, in view of the obvious importance of this development, to make it the subject of separate discussion in this postscript and update the earlier analysis in this context.

13.1 Why a productivity scheme?

We saw in Chapter 9 that the possibility of introducing a productivity scheme which would provide for additional payment on top of the fixed NPLA rate had been considered in coalmining as far back as 1971. The Wilberforce Report of 1972 recommended that a wage incentive scheme be introduced but negotiations proved abortive, as the NCB's and NUM's views on what was needed were so different. The Pay Board Report of 1974 endorsed the view of the Wilberforce Enquiry and the Report of the Tripartite Examination, consisting of representatives of government, NCB and NUM, in the same year emphasised the importance of an effective scheme being introduced. We also outlined in Chapter 8 how, despite getting the NCB to accept several amendments to a pit-based productivity scheme proposed in September 1974, the NUM recommended, successfully, that the scheme be rejected. In its place a national scheme was implemented, which was subsequently withdrawn.

The most important background factor to these discussions was the static and often declining productivity (OMS) in the first half of the 1970s and, although it was not wholly dependent on productivity, the need to halt the falling trend of coal output. The NCB also realised that an incentive scheme biased towards coalface production workers would provide a useful, and possibly less contentious, method of increasing the internal pay differential between them and other miners than would an attempt to do so in the annual wage round.

Of ever-increasing concern to both the NUM and NCB, however, was the fact that because of wage drift in the private sector, the miners, as a group, were gradually losing their hard-earned external relative wage advantage over other workers under successive phases of government incomes policy up to mid-1977. This posed the distinct possibility of another confrontation between the NUM and, this time, a Labour government. And that would have been even more likely after August 1977 if the miners had stood back and, unlike many other important groups, failed to take advantage of the opportunity to conclude a 'self-financing productivity deal' allowed under Phase Three, and subsequently extended into Phase IV, of the government's incomes policy. Coal Board estimates, based on the October Earnings' Enquiry, put miners' average weekly earnings, in 1977, at about 5 per cent above those of adult males in manufacturing compared with the 20 per cent advantage they had enjoyed on average in 1975.[1] This, then, was the background to what was to prove a bitter debate within the NUM on wage incentives towards the end of 1977, which eventually resulted in an agreement in November of that year.

13.2 The scheme

The productivity scheme proposed by the NCB was modified only slightly in negotiation. Unfortunately, the same could not be said for the *way* in which the scheme was introduced and, as with the NPLA, the method of introduction has possibly proved as relevant to its initial impact as the nature of the agreement itself. Let us deal first with the agreement.

The object of the scheme was 'to provide improvements in real wages based on improvements in productivity'.[2] To this effect the NUM's National Executive Committee decided to allow NUM areas to conclude incentive schemes at local level. This then is the most important principle of the agreement (although it is a national agreement, only the framework is nationally determined): the details are decided at pit level by area agreement.

The incentive scheme distinguishes between incentive and bonus payments. Incentive pay proper is payable to men working under an installation agreement. An installation for the purposes of the agreement means 'any coalface, development drivage (where the planned advance is 30 yards or more) and any other work place where work is performed within Grade A under the Third Daywage Structure Agreement which is exactly measurable and repetitive'.[3]

For the purpose of assessment 'standard tasks' and 'standard manning' are agreed by conventional method study techniques. Having agreed the standard task,[4] 'basic task' is defined as 75 per cent of the standard task. Incentive pay is calculated weekly, by subtracting the basic task per manshift from the actual task achieved, and multiplying the difference in lineal yards, adjusted for the depth of cut, between the standard task per manshift and the basic task per manshift. This *incentive pay* should only be paid to people working in those places (previously defined) covered by installation agreements. The remainder of the colliery labour force receive *a bonus*, based on either the average incentive pay calculated for an individual colliery, or the average for an NCB or an NUM area. The colliery or NCB or NUM area bonus is paid for normal shifts actually worked according to the following proportions:

(1) 65% of the appropriate average incentive pay per manshift for any shift actually worked on installations covered by the agreement.
(2) 50% of the appropriate average incentive pay per manshift for any shift actually worked underground.
(3) 40% of the appropriate average incentive pay per manshift for any shift actually worked on the surface.

13.3 The split within the NUM

We predicted, in Chapter 8, a sharp division within the NUM over what is essentially a pit-based incentive scheme; we were not disappointed. The idea of local incentive schemes was narrowly rejected at the NUM's Annual Conference in July. Instead, the NUM's negotiators were instructed to go ahead with a national pay claim, which if successful would have raised power-loading rates by over 90 per cent at a time when the government was attempting to restrict annual increases to 10 per cent under Phase III of its incomes policy. Opposition to a return to wage incentives, and support for the national pay claim, was led by Arthur Scargill and his Yorkshire area. The reaction of the NUM area leaders who supported the incentive proposals, which at the time were predicted by the NCB to raise average earnings in some high productivity areas by some £20 a week, was to demand a secret pithead ballot of the NUM membership. Joe Gormley, the President of the NUM, supported this course of action and his move was approved by a vote of 13–11 within the National Executive Committee. Thus began a constitutional battle within the union on the standing of Annual Conference Resolutions.

This led, in October, to the Kent miners taking the case against a pithead ballot to the High Court. In rejecting the Kent case the Vice Chancellor, Sir Robert Megamy, described the secret ballot as 'the very essence of the democratic process the Kent miners were trying to suppress'.[5] Determined as ever, the Kent Area took its case immediately to the Appeal Court. There, Lord Denning found the NEC's decision to be 'sensible and reasonable' and dismissed the Kent Area's contention that, under NUM rules, the executive should have called a special conference to discuss the latest incentive proposals, and that a secret ballot should have been called by that Conference.

Fighting on the slogan 'incentive to dig and let die',[6] Arthur Scargill launched a reported £4,000[7] advertising campaign to persuade miners to vote against the proposed productivity bonus scheme. Part of that campaign included the Yorkshire NUM buying space in the *Yorkshire Post* to advertise their case against the scheme. Just how much impact the campaign by the so-called 'Left' had on other areas is difficult to judge but, against most predictions, the pithead ballot resulted in 55¾ per cent of those voting rejecting the scheme. The full results of the ballot are set out in Table 13.1.

Although the stage seemed set for a confrontation between the miners and the Labour government, events were soon to take a different turn. All but seven areas – Yorkshire, Scotland, Derbyshire, Scottish Enginemen, South Wales, Kent, and Cumberland – had actually voted in favour of the incentive deal proposals and in some areas the vote in favour had been quite substantial. By the beginning of November reports were coming in from some areas that unofficial incentive deals were in operation at some pits.[8] Many of the rumours came from the North Derbyshire area of the NCB.[9] In response to these rumours, Nottinghamshire lodge secretaries[10] made the following recommendations to their Executive:

'(i) The executive should investigate what payments or other irregularities have been operated contrary to national agreements.

(ii) The executive should investigate the manner in which the recent ballot on the national incentive scheme was operated.

Table 13.1. *Results of pithead ballot on productivity schemes, October 1977*

Area	Total	'Yes'	(%)	'No'	(%)
Yorkshire	47,644	11,080	(23.25)	36,564	(76.75)*
Nottinghamshire	25,250	15,208	(66.00)	10,042	(34.00)
South Wales	21,572	3,643	(17.00)	17,929	(83.00)*
Officials and Staff	14,369	11,849	(82.50)	2,520	(17.50)
Durham	13,433	8,449	(63.00)	4,984	(37.00)
Scotland	13,154	2,255	(17.00)	10,899	(83.00)*
Midlands	11,182	6,019	(54.00)	5,163	(46.00)
Derbyshire	9,069	3,419	(37.75)	5,630	(62.25)*
North-Western	7,422	4,045	(54.50)	3,377	(45.50)
Northumberland	5,863	3,819	(65.00)	2,004	(35.00)
Durham Mechanics	5,036	3,806	(75.50)	1,230	(24.50)
Scottish Enginemen	4,108	1,003	(24.50)	3,105	(75.50)*
Power Group No. 1	3,973	3,016	(76.00)	957	(24.00)
Cokemen	3,427	2,043	(59.50)	1,384	(40.50)
South Derbyshire	2,687	1,912	(71.00)	775	(29.00)
Leicestershire	2,603	1,837	(70.50)	766	(29.50)
Kent	2,201	632	(28.75)	1,569	(71.25)*
Northumberland Mechs.	1,902	1,523	(80.00)	379	(20.00)
Cumberland	755	362	(48.00)	393	(52.00)*
Power Group 2	1,279	818	(64.00)	461	(36.00)
North Wales	889	597	(67.00)	292	(33.00)
Durham Enginemen	717	566	(79.00)	151	(21.00)
Total	198,535	87,901	(44.75)	110,634	(55.25)

*Area 'no' votes.

(iii) That the executive's permission be sought for any area which wishes to negotiate its own area incentive scheme.'

In the meantime reports were also coming in from several pits that the men had voted in the dark on incentives, as they had not fully understood the implications of the scheme for pay. Just what effect the last proposal of the Nottinghamshire lodge secretaries had is again difficult to gauge, but it was certainly not unwelcome to the NCB and in some NUM circles. Within four days of the proposal appearing in the press other NUM leaders had joined in the revolt against the results of the pithead ballot. By ten votes to two the executive of the NUM's Lancashire area decided to seek a scheme in line with the NCB's plan. The Lancashire NUM was joined by the executives of North Wales, South Derbyshire, Leicestershire, the Midlands and the North East, as well as Nottinghamshire. In addition, some lodges in Yorkshire were calling for another ballot in the Area in protest at what they called 'lies and propaganda'. Despite continued hostility to wage incentives in some areas, several coalfields were, by the end of 1977, operating the scheme agreed by the NUM and NCB and outlined earlier in this chapter.

13.4 PAY: INTERNAL DIFFERENTIALS, AND EXTERNAL RELATIVITIES

Incentive schemes are usually of two types. First there are what we might loosely call 'improvement schemes'. This type of scheme can be defined as one which offers extra money for output or effort in excess of what is *already* being achieved. The drawback of this type of scheme is that it tends to penalise those people who are already working at, or near, maximum efficiency. 'Equity schemes' avoid this by giving extra pay for effort or output above a uniform standard. Under this type of scheme people receive additional rewards when their performance reaches or exceeds that agreed level. This type of scheme prevents the more efficient employees from being penalised in relation to the less efficient. The coal industry's scheme falls most appropriately into this second category.

The NCB and NUM agreed that the incentive scheme would pay 94 pence per week for every one percentage point of output in excess of the basic task to be set at 75 per cent of the standard task. Thus, if output was 25 percentage points (i.e. $33\frac{1}{3}$ per cent) above the basic task, men covered by an installation agreement would get £23.50 per week (i.e. 25×94 pence). As the sum of £23.50 represented an increase of $33\frac{1}{3}$ per cent on the weekly NPLA rate, percentage increases in output above the basic task were to be rewarded by an equivalent percentage increase in pay. In accordance with the principles of an 'equity scheme', any group of miners whose output was already in excess of the agreed basic task when an installation agreement came into effect would immediately receive a wage increase. A failure to fully appreciate this feature of the scheme seems to have been partly responsible for it being rejected at some pits.

The mechanics of the incentive scheme have led to the accusation that it is a soft option. Most of its features, however, are quite common to incentive schemes in other parts of British industry. It was certainly a fact, though, that in some instances the NCB would have to pay more for the same results. However, taking the industry as a whole, the NCB claim that when the scheme was introduced average installation performance was at a level which corresponded to the base line of 75 per cent of the average standards agreed at which incentive pay would begin to accrue. The trick then, if the scheme was not to increase average unit labour costs, was to raise labour efficiency in pits where tasks being achieved were substantially below agreed basic tasks. Before we explore this aspect of the scheme in any depth, let us look at the effects of the scheme on miners' pay.

For the reasons we have outlined earlier, the incentive scheme was not introduced into every pit at the same time. Pits in South Derbyshire, Nottinghamshire, Warwickshire and Staffordshire were among the first to conclude local installation agreements. These areas were followed by the North East and Scotland. Predictably perhaps, given the initial strong reaction to the notion of local wage incentives, Yorkshire was the last area to conclude incentive agreements (in April). It was not until May 1978, therefore, that the scheme was operating throughout the industry.

Earnings from installation agreements vary enormously between coalfields and, within fields, between pits. Average incentive pay for September 1978, for example, varied from £12.90 a week in South Wales to £37.29 in North Wales.[11] There are only two pits in North Wales, however: in one the average incentive pay for September was £54.80, while in the other the average was £19.77. Another example of the large variation in incentive pay within an area comes from Doncaster, where, although the average area incentive pay was £20.06 in September 1978, it varied from £6.55 at one pit to £43.90 at another. For the industry as a whole, average incentive pay for men operating under installation agreements was £20.00 a week in September 1978.

We explained earlier how men whose work was not covered by an installation agreement were to receive a *bonus* fixed as an agreed proportion of the average *incentive pay* of a colliery or NCB or NUM area. In the majority of cases men chose to have their bonus determined on a colliery rather than an area basis. Average bonus payments in September 1978 were £10.50 for men working underground other than at the coalface and £8.40 for men working at the pit surface.

The productivity scheme had the immediate effect of increasing internal wage differentials thus reversing the trend of the previous twenty years. Table 13.2, which gives the EMS of men working at the coalface, elsewhere underground, and at the pit surface for some months before and after the agreements came into effect, shows that the internal differential between face and development workers (grade U1) and surface daywagemen excluding craftsmen (grades S1-S6) increased by 13 percentage points between October 1977 and October 1978, whereas the differential between non-craft daywagemen working underground behind the coalface (grades U2-U7) and

Table 13.2. *EMS and internal wage differentials 1977–79*

	Earnings per manshift				Differentials			
	(Grades)[a]				U1 / S1-S6	U1 / U4-U7	U1 / U2-U3	U4-7 / S1-S6
Date	U1	U2-U3	U4-U7	S1-S6				
October 1977	15.71	13.60	12.89	11.78	133.4	121.9	115.5	109.4
April 1978	20.65	16.97	16.13	14.33	144.1	128.0	121.7	112.6
October 1978	21.21	17.30	16.51	14.48	146.5	128.5	122.6	114.0
April 1979	24.32	19.63	18.88	16.72	145.5	128.8	123.9	113.0
Percentage increase								
Oct '77–Oct '78	35.0	27.2	28.0	22.9				
Oct '78–Apr '79	14.7	13.5	14.4	15.5				

[a]U1 NPLA and Third Daywage Structure grade A
U2-3 Third Daywage Structure grade B and Third Daywage Structure grade C
U4-7 Other daywagemen behind the coalface, excluding craftsmen
S1-6 Surface daywagemen excluding craftsmen
Source: NCB, Supplementary Form F (April and October Earnings Enquiry)

non-craft surface daywagemen increased by 8 percentage points on average.[12]

These internal wage adjustments were accompanied by what, for even the lowest-paid surface daywagemen, were fairly generous wage increases emanating from the productivity agreement. The agreement was negotiated during a period when the government incomes policy norm was a 10 per cent increase in earnings. After much argument the miners settled nationally, in March, within the 10 per cent guidelines. However, there can be no doubt that their acceptance of the Phase III guidelines owed much to the productivity scheme which, together with the national award, raised the average EMS of non-craft surface daywagemen by nearly 23 per cent and the average EMS of face and development workers and Third Daywage Structure grade A by 35 per cent between October 1977 and October 1978.

In the country as a whole average earnings in manufacturing increased by approximately 15 per cent during Phase III of the government's incomes policy (August 1977–August 1978). The productivity agreement in coalmining thus had the effect of checking the gradual deterioration which had occurred in the miners' external wage relativity after 1975. Figure 13.1 recaps on, and updates, the trend in the miners' external wage relativity since nationalisation which was outlined in Chapter 8. It shows that by April 1978,

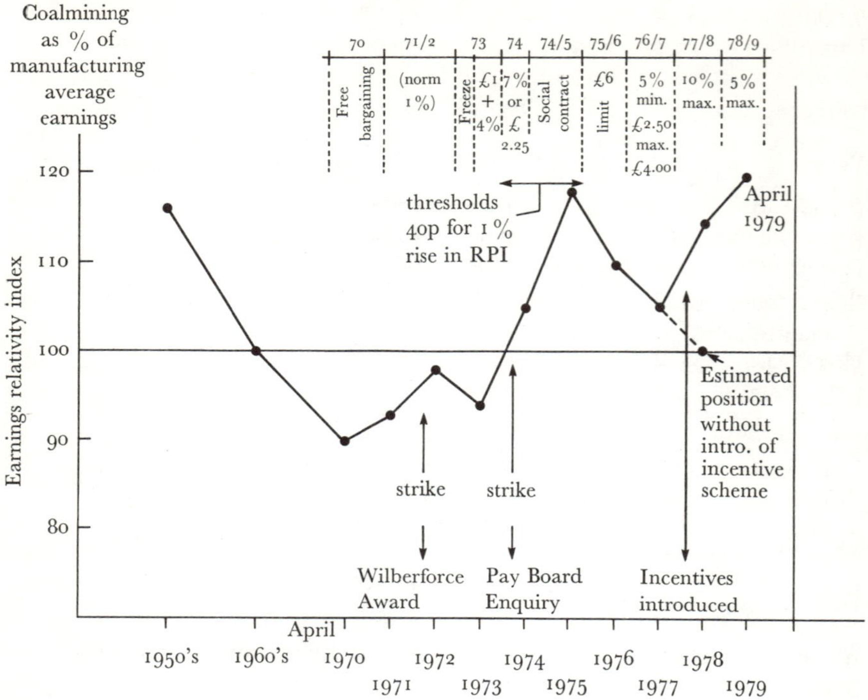

Figure 13.1 Earnings relativities since the 1950s

without the productivity bonus, the miners' external earnings relativity would (on the basis of earnings trends in mining and manufacturing) have fallen back to the position just before the national mining strike of 1974. The productivity bonus effectively raised the miners' earnings relativity some 15 percentage points above the level which it would otherwise have been, and to a position similar to that of the late 1950s. Once again the mining industry went to the top of the UK earnings league.

The miners improved still further on their relative wage position through their national wage award of March 1979. That award was a fairly complicated package designed to avoid embarrassing a Labour government, which was struggling to save an incomes policy in an election year. The agreement provided for an average increase of about 9 per cent on the basic wage bill, although the total package, which included raising the incentive payment for achieving standard task performance to £26.50 a week, was substantially in excess of this figure. Joe Gormley, the miners' President, urged that the 9 per cent basic wage increase be biased in monetary terms, towards face and development workers; instead, the NUM members opted for a flat-rate increase of £6.50 for all mineworkers.

This agreement helps to explain the reduction in the internal differential between miners working at the pit surface and those in U1 jobs working underground, by April 1979, which is observed in Table 13.2. However, other factors were operating to the same effect. First, percentage increases in the incentive pay element had been smaller than the percentage increase in basic wage rates. Second, there had been a sharp increase in the incidence of overtime working among the lower-paid surface daywagemen. The impact of these factors on internal differentials was partially offset, however, by a 'knock-on' effect of higher incentive earnings at the coalface to other U1 jobs which has served to raise U1 EMS. An element of 'drift' in incentive earnings has operated in the same direction.

13.5 Output and productivity

Thus far the productivity scheme can be seen as having had three major effects. First, it restored miners as a whole to the top of the industrial earnings league. Second, and as a direct consequence, it probably averted a national strike in 1978 and, because of its inevitably divisive effects within the NUM, may have diminished the likelihood of a national strike in the near future. Third, it produced the widening of internal differentials now appreciated as essential to encourage miners to train for, and move on to, the coalface, although this trend is now in danger of being reversed by national agreements, and, perhaps inadequate wage control procedures. The main explicit object of the 1977 agreement, however, was to maintain, if not actually increase output and productivity. To what extent, therefore, has the agreement been successful in this respect?

Some impression of the wage incentive schemes' initial impact can be obtained from Table 13.3. Productivity increased quite sub-

Table 13.3. *Coal output, productivity and employment 1975–78*

	Coal output deep mined (000 tonnes)	Productivity (OMS tonnes) Overall	Face	Wage earners on colliery books (000s)
1975	117.4	2.28	7.92	245
1976	110.3	2.23	7.75	241
1977	107.1	2.18	7.77	239
1978	107.5	2.25	8.53	232

Source: Department of Energy, *Energy Trends*.

stantially during 1978: comparing the years 1977 and 1978, overall OMS increased by 3.2 per cent, while face OMS increased by 9.8 per cent. These productivity improvements were sufficient to ensure a slow expansion of coal output, while the colliery labour force was reduced by 7,000.

If one is to measure the true impact of the incentive payment schemes on output and productivity, however, one should do more than point to the increase in output and productivity, which occurred during 1978. One of the main purposes of the agreements was to halt, what the Coal Board estimated as likely to be a further substantial *decline* in both productivity and output, without such agreements. Taking the years 1975–77, for example, overall OMS was declining by approximately 2.3 per cent per annum. If one adds this to the 3.2 per cent increase in overall OMS in 1978, it is probable that the true impact of the incentive schemes on overall productivity was in the region of 5–6 per cent. By any standards this was quite a significant achievement.

The question now remains as to whether these achievements will be sustained, or even improved upon. We can speculate on this question with the help of the quarterly productivity data presented in Table 13.4. The data show that both face and overall productivity displayed a strong upward trend throughout 1978. In the quarter October–December 1978, for example, overall OMS was 6 per cent

Table 13.4. *Quarterly productivity comparisons 1977–79*

Period	Overall OMS (tonnes)	% increase	Face OMS (tonnes)	% increase
January–March 1977	2.27		7.93	
January–March 1978	2.30	1.3	8.47	6.8
April–June 1977	2.22		7.94	
April–June 1978	2.25	1.7	8.65	8.9
July–September 1977	2.02		7.44	
July–September 1978	2.09	3.5	8.24	10.8
October–December 1977	2.18		7.70	
October–December 1978	2.31	6.0	8.63	12.1
January–March 1978	2.30		8.47	
January–March 1979	2.27	– 1.3	8.55	1.0
January–March 1977	2.27		7.93	
January–March 1979	2.27	0	8.55	7.8

Source: CSO, *Monthly Digest of Statistics*.

above that for the same quarter in 1977, and face OMS had increased by 12 per cent over the same period.

In the first quarter of 1979, however, overall OMS was some 1.3 per cent down on the same quarter of 1978 and was, in fact, no higher than in the same quarter of 1977. Occurring as it did against a background of falling employment, this disappointing trend in overall productivity contributed to deep-mined coal production being some 2 per cent lower in the first quarter of 1979 as compared with the first quarter of 1978. However, although the rate of increase in face OMS slowed down dramatically in the first quarter of 1979, the absolute level of face OMS was still some 7.8 per cent above the level of two years previously. How is it then, that face OMS first rose at a much faster rate than overall OMS, and more recently, overall OMS has fallen while face OMS has been sustained at its improved level? These questions assume a particular importance in view of the fact that incentive and bonus payments are governed by *face*, rather than overall productivity improvements.

A change in the allocation of labour between the coalface and other parts of the pit will clearly affect face productivity much more in the short term than it will overall productivity, although the efficiency of labour allocation might be expected to affect the latter in the medium and longer term. Labour allocation will depend on a number of factors of which the most important are: recruitment flows, the availability of face-trained labour, the need for development work, as well as the individual effort bargains which influence manning norms, and overtime working. If we refer back to Table 13.3, it is apparent that face productivity declined less fast over the period 1975–77 than overall OMS. This can be explained partly by the influx of new recruits to the industry. These recruits had to be trained at the pit surface and *behind* the coalface before they could be deployed to existing or new coalfaces. This increased the proportion of shifts worked away from the coalface. Assuming everything else to be equal, therefore, overall productivity would inevitably be depressed in the medium term relative to face productivity, which would be affected only by the extent to which the influx of labour improved services to the coalface.

By the beginning of 1978, when the installation agreements were gradually coming into effect, the tendency outlined above was moving into reverse: labour recruited in the previous two years was now trained and thus able to be used more effectively with a beneficial effect on both face output and productivity. In addition, though, there was now an incentive (as under the old price lists) for face teams to shed labour and this, together with the general increased availability of suitable labour, meant that an increasing amount of manpower could be deployed to new faces and previously neglected development work. At one and the same time, therefore, there was a positive effect on overall productivity (and output) but a much larger increase in face productivity as labour was deployed elsewhere.

So much then for the way in which discrepancies between face and overall OMS could be brought about by the legitimate and intended operation of installation agreements. There are, however, ways of artificially raising face OMS to bolster incentive and bonus earnings, which have no positive effect on output. As bonuses are fixed as a proportion of the incentive pay earned on installation agreements, it is in the interest of all miners at a pit working such an agreement to minimise the number of shifts recorded at the coalface.

Incentive payments can be increased if shifts actually worked at the coalface can be booked to work in other parts of the pit. Again, a failure to record overtime shifts actually worked by miners on installation agreements would also raise both incentive and bonus pay. For the amount of time worked, incentive and bonus pay is, in many instances, a more lucrative alternative to overtime premium pay, and, of course, more people benefit from such a substitution than the face teams actually working the overtime. However, whereas the booking of shifts worked at the coalface to other places of work will tend to enhance face OMS at the cost of OMS overall, the failure to actually record some of the shifts worked in overtime at all, will raise both OMS statistics. This writer has no way of estimating the impact of what are but two examples of wage malpractices on productivity, and thus pay. By the very nature of the problems inherent in operating new agreements, however, one can be fairly confident that both have been used to some effect.

While the practices outlined above will certainly tend to distort the relationship between face and overall OMS, and in some cases even 'artificially' raise overall OMS, it is exceedingly doubtful if they in any way account for the absolute decline in deep-mined coal production or even overall productivity in the first quarter of 1979.

At any given level of face productivity, total output and overall productivity will be adversely affected if there is a reduction in the number of coalfaces in operation, or if there is a shortage of labour to man existing faces. While the cutback on recruitment towards the end of 1978 had a positive effect on overall productivity in the short term, the evidence would suggest that it depleted the pool of 'green labour' available for tasks behind the coalface. This deficiency would appear to have been met by redeploying men from the coalface, thus leaving a shortage of manpower for both existing and new faces. A study conducted by the Coal Board in the second quarter of 1979, for example, revealed that approximately 13.0 per cent of U1 graded men eligible for incentive pay were working on one-U1 graded jobs behind the coalface. While this finding means that more men working behind the coalface are qualifying for incentive pay than should be the case, it is primarily indicative of the shortage of labour available for tasks elsewhere underground. In some areas of Yorkshire the deficiency has proved to be in excess of 20.0 per cent.

The deployment of men between the coalface and operations elsewhere underground is now deserving of very serious examination, as is the current recruitment policy of the NCB, if these somewhat disconcerting trends are to be reversed. The position in April 1979 would appear to be that the NCB is paying more for approximately the same amount of coal as it was getting in 1977.

13.6 Stoppages and restrictions of work

We saw in Chapters 8 and 10 that the continued decline in the number of work stoppages and restrictions in coalmining after the NPLA in 1966 was accompanied by, and partly due to, a very sharp reduction in the number of disputes over pay as piecework gradually gave way to fixed daywages at the coalface and jobs elsewhere underground. While that trend continued into the late 1960s and early 1970s, the number of working days lost over pay increased quite substantially, as miners took an increasingly entrenched stand over the relative deterioration in their living standards which accompanied the transition from wage incentives to fixed day rates.

The incentive scheme drawn up by the NCB and NUM was fairly complicated and it was predictable that it would result in an increased incidence of disputation and, in turn, an increased number of stoppages and restrictions of work. Although it may be too early in the life of incentive agreements to draw firm conclusions about their impact on industrial relations, a comparison of the number of work stoppages and restrictions during the six months ending September 1978 with the same period in 1977, given in table 13.5, does conform to our prediction. Thus nearly 70 per cent of the increase in industrial disputes resulting in stoppages and restrictions of work over the period was attributable to disputes over pay, and disputes over bonus payments in particular. The number of manshifts lost due to stoppages and restrictions of work rose between the two periods from 36,130 to 132,961.[13] Of that increase 75 per cent was due to stoppages and restrictions of work over pay. The most common cause of disputes

Table 13.5. *Stoppages and restrictions of work 1977–78*

	Disputes Six months ending September 1977		Disputes Six months ending September 1978	
Cause of dispute	No.	%[a]	No.	%[a]
Allowances and bonuses	44	13.7	137	29.2
Wages and price lists (other than above)	12	3.8	23	4.9
Refusal of alternative work	40	12.5	49	10.4
Refusal to perform work left over from previous shift	10	3.1	11	2.3
Refusal to repair after mechanical breakdown	4	1.2	2	0.4
Methods of working or organisation	89	27.7	70	14.9
Bad working conditions	29	9.0	29	6.2
Objections to officials	25	7.8	12	2.6
Sympathy with men dismissed or reprimanded	9	2.8	16	3.4
Retaliatory	18	5.6	69	14.7
Miscellaneous	41	12.8	51	10.9
Total	321	100.0	469	100.0

[a]The figures do not sum precisely to 100 due to rounding.
Source: National Coal Board.

under the allowances heading was over manning standards on installation agreements.

Against this fairly predictable increase in stoppages and work restrictions during the early phase of installation agreements, one must weigh the very high probability that a national mining strike was averted as a consequence of the return to local bargaining. Furthermore, the inevitable fragmentary effect which the fairly large disparity in pay which now exists between coalfields and the pits within them must have within the NUM, has probably made such an eventuality less likely in the future. It remains to be seen whether the NCB can exploit this potential source of weakness within the NUM by attempting to get the union to accept that the bulk of future wage increases should be based on local incentives and bonuses, rather than on standard wage rates. Certainly the incentive agreements afford the opportunity of diffusing the pay issue.

13.7 ACCIDENTS

Several opponents of wage incentives predicted that the scheme would result in more pit accidents. The accident rate in coalmining improved substantially in the late 1960s, and 1970s, as better and safer technology was introduced into the pits; between 1965 and 1971 the accident rate underground per 100,000 manshifts worked fell from 243.5 to 159.0.[14] By the year ending March 1978 it had fallen even further, to 112.1. Table 13.6, which compares various accident rates by cause for the 39 weeks ending December 1978 – the period when installation agreements were operating in virtually all pits – with the same period in 1977, does not enable one to come to any hard and fast conclusions as to the effect of a return to wage incentives on accidents.

Overall, accident rates continued to decline after the introduction of incentive agreements, and of the seven causal categories only three – falling objects, haulage and transport behind the coalface, and stumbling and falling – increased and then only very slightly.

Data on deaths are, superficially at least, rather less favourable: the number of deaths rose by 4 underground and 9 at the surface. More alarming from our viewpoint perhaps is the rise caused by falls of ground at the coalface (+5) and by haulage and transport behind the coalface (+5).

It would be dangerous indeed to attribute the increase in deaths wholly to men taking greater risks in the quest for more money. Deaths on haulage and transport had risen quite sharply before the incentive schemes were introduced. Thus in the year ending March 1978 deaths from this cause were 11 more than in the previous year.[15] This trend has been of very serious concern to the NCB and NUM but so far rigorous analysis of haulage accidents has failed to reveal clear explanations for their increase. Over a third of the fatalities on haulage and transport occurred to men in the age group 56–60 and the second highest fatality rate was amongst men in the group aged between 36 and 40 years. These findings might suggest that as men get older they also get more careless. It might also indicate that older men, and particularly those who have recently returned to the pit, are not as familiar with developments in haulage systems as they should be. Accidents were found to be most frequent on Mondays when absence is usually at its highest. This further suggests that redeployment of labour without the essential experience and training to cover for labour shortages behind the coalface may be a factor in the upward trend of haulage accidents. We can only hope that the trend in haulage accidents will be reversed as training is intensified and

Table 13.6. *Accidents and deaths at NCB mines 1977–78 (39 weeks ended 23 December 1978 and comparison with similar period 1977)*

	Fatal		Serious Reportable		Total over 3 days incapacity (including fatal and serious reportable)	
	Number	Change on 1977	Rate per 100,000 manshifts	Change on 1977	Rate per 100,000 manshifts	Change on 1977
Falls of ground at face	8	+5	0.02	−0.05	12.9	−0.5
Falls of ground in roadways	3	—	0.04	0.02	3.5	−0.1
Falling objects	0	—	0.09	−0.02	12.6	+0.1
Haulage and transport at face	0	−1	0.04	0.00	0.5	−0.1
Haulage and transport elsewhere underground	20	+5	0.03	0.01	5.9	+0.2
Machinery – face	2	−1	0.08	−0.03	2.3	−0.3
Machinery – elsewhere underground	0	−1	0.05	0.02	1.1	−0.1
Tools and appliances	0	—	0.04	0.02	6.7	−0.5
Haulage supplies	0	−1	0.03	−0.02	22.9	—
Stumbling or falling	1	—	0.15	0.00	29.4	+0.1
Other	0	−2	−0.02	−0.02	12.4	−0.1
Total underground	34	+4	1.03	−0.08	110.1	−1.7
Surface	14	+9	0.44	0.01	42.4	−1.1
Total	48	+13	0.90	−0.06	95.6	−1.7

Source: National Coal Board.

men made increasingly aware of the dangers associated with modern haulage and transport systems.

A similar set of factors to those discussed in the previous paragraph may also be responsible for the increase in deaths at the coalface, although the evidence is less convincing. The incentive scheme offers an inducement to reduce manning at the coalface. It is often the case that the slower but perhaps more careful men are the ones who are squeezed out by this process of de-manning. Where this occurs the risk element increases proportionately. However, if it were the only factor at work we might have expected the accident rate to increase rather than decline. Thus, while there are certainly some aspects of the incentive agreements which could lead to a greater element of risk, our evidence is not wholly conclusive. Certainly, though, the situation must be kept under careful surveillance.

13.8 Summary and conclusions

Having moved gradually away from incentive payment schemes in the mid 1960s and early 1970s the coalmining industry reversed that decision in 1978. The object of the scheme was to arrest the declining trend of output and productivity, raise the miners' external pay relativity without resort to a national confrontation on pay during a period of central incomes restraint, and insofar as the NCB was concerned widen internal pay differentials, which, at their prevailing levels, provided insufficient incentive for men to move on to the coalface. While our analysis would suggest that each of these objectives was achieved in the short run, there is a danger that short-run productivity improvements might be squandered by inadequate manpower policies and wage control procedures.

However, the main conclusion of the study remains that whatever preconceived ideas organisations may have on pay, their systems and policies will ultimately have to adapt to the environment in which they find themselves operating. While centralised pay bargaining remains an important force in coalmining, it has become more flexible, so as to accommodate the strains imposed upon it by individual and collective responses to incomes policy, structural change and the needs of a more favourable market environment. It remains to be seen whether this trend will be reversed again at a later date.

NOTES

NOTES TO CHAPTER 1

1 One million tons of petroleum is equivalent to 1.7 million tons of coal – see Department of Trade and Industry, *United Kingdom Energy Statistics*, 1973, p. 180, HMSO.

2 *Digest of Energy Statistics*.

3 *Fuel Policy*, Cmnd. 3438 (1967).

4 For a detailed description of coal types, see A. Beacham, *The Coal Industry*, pp. 108–55, in *The Structure of British Industry*, Vol. 1, edited by D. Burn, NIESR, CUP, 1958.

5 This description was substantially accurate until the early 1970s by which time stable holes were largely machine cut. By the mid 1970s, two power-loading machines cutting in opposite directions were deployed on most coalfaces. This eliminated the need for stable holes.

6 Report of the Advisory Committee on the Technical Organisation of the Coal Industry, 1945.

7 Report of the Advisory Committee on Organisation (Fleck Report), NCB, February 1955.

8 National Union of Mineworkers, *Rule Book*.

9 *Ibid*.

10 *Ibid*.

10[a] These agreements were superseded by a demarcation agreement in 1970 which gave NACODS exclusive negotiating rights for officials and COSA similar rights for WPIs.

11 A change in wage district boundaries followed the management reorganisation of March 1967.

12 Coal Industry Commission, *Report* (1925), Volume 1. Report Cmd. 2600, page 132.

13 Sliding scales persisted until 1902 in South Wales and were not abandoned in Northumberland until the advent of government control of the industry in 1917.

14 Wages still tended to follow coal price fluctuations and greater wage stability in areas covered by conciliation boards probably owed more to the greater stability of coal prices in those areas than to the boards per se.

15 For an account of this, see R. Page Arnot, *The Miners; Years of Struggle*, Allen and Unwin, pp. 161–4.

16 Where this did not provide a subsistence wage for low-paid daywagemen it was to be made up by allowances – the scale of allowances was to be determined by the District Board or the individual chairman if the Board could not agree.

17 Other recommendations were an extension of payments-by-result schemes, statutory profit sharing and cooperative selling schemes.

18 Shift times were generally increased to eight hours underground and the week lengthened to 48 or 49 hours for surface workers.

19 Cmd. 6364, 1942.

NOTES TO CHAPTER 2

1 See Harold Wilson, *New Deal for Coal*, London 1945, for an account of progress on the wages front during the war years.
2 NCB, *Memorandum*, Part 2, pp. 106–61, 1946.
3 *Ibid.*
4 NCB, *Annual Report and Statement of Accounts*, 1947, p. 118.
5 NUM, *Annual Conference*, Report and Proceedings, 1946, pp. 53–67.
6 NUM,*Annual Conference*, Report and Proceedings, 1948, pp. 63–7; especially the speech by Abe Moffat.
7 NUM, *Annual Conference*, Report and Proceedings, 1947, p. 66.
8 General Secretary of the NUM.
9 See debates on wages in NUM, *Annual Conference*, Report and Proceedings, 1946–48.
10 NCB, *Report and Accounts*, 1948.
11 For a report on these negotiations, see NCB, *Report*, 1947.
12 NCB, *Report*, 1947.
13 NUM, Minutes of National Executive Committee, 1946, and NUM, *Annual Conference*, Report and Proceedings, 1947, p. 69.
14 NUM, *Annual Conference*, Report and Proceedings, 1947, p. 70.
15 G. B. Baldwin, *Beyond Nationalisation*, p. 126.
16 This tends to confirm the proposition of Arthur Horner that 'as a result of what happened in 1944 – and as a result of the 16% addition to piece-rate earnings – pieceworkers are earning wages far in excess of the daywagemen, and far beyond that which previously they had the ability to earn', NUM, *Annual Conference*, Report and Proceedings, 1947, pp. 69–70.
17 See Chapter 1.
18 NCB, *Annual Report*, 1948.
19 For an account of the Tribunal's findings, see NCB, *Memorandum*, Part 7, p. 462.
20 NCB, *Annual Report*, 1956.
21 There was also a special grade for winding engineers.
22 Although the grading of jobs was not decided using a formal points rating system or similar technique, the exercise was a form of job evaluation based on practical experience of the work involved in each job.
23 The full text of the agreement can be found in NCB, *Memorandum*, Part 10, 1956.
24 *Ibid.*
25 W. H. Sales and J. L. Davies, 'Introducing a New Wage Structure into Coal Mining', *Bulletin of the Oxford University Institute of Statistics*, Vol. 19, No. 3, 1957, p. 201.
26 This figure is an average of the wage increases agreed since 1955.
27 NUM, *Annual Conference*, Report and Proceedings, 1955.
28 *Ibid.*, p. 70.
29 NUM, *Annual Conference*, Report and Proceedings, 1955, p. 73.
30 *Ibid.*
31 Joint Working Party Papers, NCB/NUM, 1952–54.
32 NCB, *Annual Report*, 1947.
33 At this time regular overtime was being worked. It is important to note also that, because they were paid by the week, the Five-Day-Week Agreement did not apply to overmen, whereas it did apply to deputies.
34 NCB, *Annual Report*, 1952.
35 NCB, *Annual Report*, 1947.

36 See NCB, *Memorandum*, Part 11, 1956, pp. 97–118, for a full description of the agreement.

37 *Ibid.*

NOTES TO CHAPTER 3

1 NUM, *Annual Conference,* Report and Proceedings, 1946.

2 It is as well to point out here that, at the coalface, the piecework system we describe in this section has been substantially amended as new techniques of coal getting (notably power loading) have been introduced. And, although this description is still applicable to the piecerate system as operated on the few coalfaces mined by conventional techniques and to 'elsewhere underground' pieceworking, I have chosen, to write this section in the past tense.

3 NCB, Working Party Papers, 1957.

4 Although the form of the agreements drawn up for power-loading machines were different in many respects from those for pieceworkers on machine-cut faces, the allowance system, as described here, was similar.

5 The data are taken from Table 3.5 using the mid-point of the class interval and 22.5 and 112.5 for the two open class intervals.

6 *The Coal Industry Nationalisation Act*, 1946, Section 2(3).

7 The survey took two years to complete. During that time 92,900 forms had been dispatched as well as 210 notes for guidance. The reports ran to 116 volumes, occupied 38 cubic feet of shelf space, weighed nearly 11 cwts and covered 173,352 pieceworkers. The programmes had cost over £4,000, 120 hours of computer time had been used and 237,766 cards had been punched. The clerical time used for the survey was estimated at 6 man years and, if everything was included, it had cost an estimated £2 millions. (These estimates are those given by the NCB's planning group.)

8 Figures in this section are derived from the Working Party Reports on the Revision of Pieceworkers' Wage Structure, 1957.

9 See the next chapter.

NOTES TO CHAPTER 4

1 NCB, *Memorandum of Agreements*, Part 2, 20 March 1940 to 31 July 1946, pp. 129–31.

2 For a report of the discussion, see NUM, *Annual Conference*, Report and Proceedings, 1958, pp. 107–14.

3 *Ibid.*, p. 109.

4 *Ibid.*, p. 108.

5 *Ibid.*, p. 111.

6 See NUM, *Annual Conference*, Report and Proceedings, 1959, p. 26.

7 NCB, *Report and Accounts*, 1963–64, p. 17, paras. 83–8.

8 See NCB, *Memorandum of Agreements*, Part 20, 1 January to 31 December 1965, pp. 860–6, for the full text of the Agreement. Also NCB, *Annual Report*, 1965–66, for the background to it.

9 NUM, *Annual Conference*, Report and Proceedings, 1965, p. 21.

10 A copy of the agreement can be found in NCB, *Memorandum of Agreements* Part 21, 1 January to 31 December 1966, pp. 56–61.

11 Clause 9, of NPLA.

12 Clause 8, of NPLA.

13 The five-day-week bonus had previously varied from 20–16% depending on whether power-loading teams were treated as pieceworkers or taskworkers.
14 Clause 10–13.
15 Clause 4.
16 Clause 4–5.
17 Clause 6.
18 Clause 23.
19 See definition of F69 Power-Loader Man, on p. 3 of the agreement.
20 Clause 3.
21 NCB, *Memorandum of Agreements*, 1946.
22 Writer's emphasis.
23 See, NCB, *Annual Report*, 1966–67.
24 Royal Commission on Trade Unions & Employers Association, *Minutes of Evidence 4*, Tuesday, 16 November 1965, Witness, National Coal Board, London, HMSO, 1966, p. 164.
25 Report No. 12, National Board for Prices and Incomes, *Coal Prices*, London, HMSO February 1966, p. 14.
26 National Board for Prices and Incomes, *Coal Prices*.
27 The proceedings of the conference are reported in NUM, *Annual Conference*, Report and Proceedings, 1966.
28 *Ibid.*
29 See NCB, *Annual Report*, 1966–67, pp. 36–7.
30 R. G. Searle-Barnes, *Pay and Productivity Bargaining*, Manchester University Press, 1969.
31 NUM, *Annual Conference*, Report and Proceedings, 1968.
32 NCB, South Midlands, Survey of Task Rates, 1967.
33 This account of national negotiations is based on NCB, *Memorandum of Agreements*, Parts 14–16, 1969–71.
34 The shift rates so defined were to be calculated exclusive of overtime premium, five-day-week bonus, and all allowances payable under the 1966 NPLA agreement.
35 Information supplied by National Coal Board.
36 When the different rates of five-day-week bonus are included, the addition to the base shift rate of *each pay group* is such as to yield the same shift rate, i.e. in the first pay group, 86/9 + 4/- + 20% is equal to 89/9d + 4/2d + 16%.

NOTES TO CHAPTER 5

1 NUM, *Annual Conference*, Report and Proceedings, 1956, p. 236.
2 NCB, *Memorandum of Agreements*, 1957, pp. 385–6.
3 This table is derived from wage dispersion data supplied by the NCB.
4 From NCB, *Wage Surveys*.
5 NUM, *Annual Conference*, Report and Proceedings, 1956, p. 244.
6 NUM, *Annual Conference*, Report and Proceedings, 1955.
7 NCB, *Memorandum of Agreements*, 1959, p. 803.
8 NCB, *Memorandum of Agreements*, 1959, p. 805.
9 NCB, *Memorandum of Agreements*, 1959, pp. 804–5.
10 NCB, *Memorandum of Agreements*, 1959, p. 807.
11 W. H. Sales and J. L. Davies, 'Introducing a New Wage Structure into Coalmining', *Bulletin of the Oxford University of Statistics*, Vol. 19, No. 3, 1957.
12 Writer's emphasis.

13 NCB, *Memorandum of Agreements*, 1957, pp. 323–4.
14 This section is based on discussion with national officials of the NCB and NUM.
15 NCB.
16 NUM, *Annual Conference*, Report and Proceedings, 1958, p. 86.
17 *Ibid.*
18 NUM, *Annual Conference*, Report and Proceedings, 1969, p. 146.
19 For details of these negotiations, see National Union of Mineworkers, *Annual Conference*, Reports and Proceedings, 1970, pp. 70, 78, 478, 484 and 567.
20 See NCB, *Annual Report*, 1957, p. 45.

NOTES TO CHAPTER 6

1 Royal Commission on Trade Unions and Employers' Associations 1965–68, *Report*, June 1968, London, HMSO Cmnd. 3623.
2 *Ibid.*, para. 1019.
3 H. A. Turner, 'Wages: Industry Rates, Workplace Rates and the Wage Drift', *Manchester School*, May 1956.
4 H. A. Turner, 'Trade Union Differentials and the Levelling of Wages', *Manchester School*, September 1952.
5 K. J. W. Alexander, 'Wages in the Coalmining Industry since Nationalisation', *Oxford Economic Papers*, Vol. 8, No. 2., June 1956.
6 H. A. Turner, 'The Disappearing Drift (Or in Defence of Turner)', *Manchester School*, May 1964.
7 H. A. Turner, 'More on the Wages Drift', *Manchester School*, January 1960.
8 H. A. Turner, 1964, p. 181.
9 M. Barratt Brown, 'Determinants of the Structure and Level of Wages in the Coalmining Industry since 1956', *Bulletin of the Oxford University Institute of Statistics*, Vol. 29, pp. 139–70, May 1967.
10 For a detailed analysis of these problems, see 'A Further Note on the Wage Drift', pp. 15–27, in National Board for Prices and Incomes, *Payment by Results Systems* Report No. 65 and Supplement, Cmnd. 3627–1 HMSO, 1968.
11 See H. A. Turner, *Manchester School*, 1960, *op. cit.*
12 H. A. Turner, *Manchester School*, 1956, *op. cit.* Also G. A. Penrice, 'Earnings and Wage Rates Since 1938' and 'Earnings and Wage Rates 1948–55' in *London and Cambridge Economic Service Bulletin*, September 1952, pp. 4–6 and December 1965, p. 11 respectively.
13 For an analysis of the effect of these changes on absenteeism, see L. J. Handy, Absenteeism and Attendance in the British Coalmining Industry since Nationalisation, *British Journal of Industrial Relations*, Vol. VI, No. 1, 1968, pp. 27–50.
14 The following figures are for May 1968.

	Age group				
	21–9	30–39	40–9	50–9	60 and over
Weekly earnings	£24.4	£24.2	£22.9	£19.9	£17.6

Source: National Coal Board.

15 The figures of EMS and average weekly earnings from this source correspond to

those published by the NCB in its Annual Reports and Accounts which have been used by other writers when examining wage movements.

16 By multiplying the underground and surface minimum wage rate by the number of shifts worked at each place of work and dividing by the total shifts worked.

17 Except Barratt Brown (1967) who, perhaps misleadingly, compared increases in shift rates and *weekly* earnings.

18 H. A. Turner, *Manchester School*, May 1964, *op. cit.*

19 See, NCB, *Annual Report*, 1951.

20 NCB, Financial Returns.

21 H. A. Turner, *Manchester School*, 1952.

NOTES TO CHAPTER 7

1 For a historical account of these districts, see Chapter 2. At nationalisation there were twenty wage districts. The three South Western districts (South Wales and Monmouthshire, Bristol and Somerset, and the Forest of Dean) were subsequently merged, however, and for the sake of continuity we have aggregated the relevant data and treated those districts as a single entity.

2 Taken from the Ministry of Power, *Statistical Returns on the Proceeds, Costs of Production, and Profits and Losses of Collieries*, 1938.

3 *Bulletin of the Oxford University Institute of Statistics*, Vol. 29.

4 NCB, *Revised Plan for Coal*, 1959.

5 Data taken from NCB *Financial Returns*.

6 National Board for Prices and Incomes, *Coal Prices*, Supplement No. 2, *Cmnd 4455–2*, HMSO, London 1971.

7 NCB, *Financial Returns*, 1967–70.

NOTES TO CHAPTER 8

1 The April enquiry was discontinued in 1970.

2 The NCB do calculate an EMS statistic for adult mineworkers (over 21), but this is given as an *annual* average before 1965: only after that year was such a statistic calculated for October and this has never been published.

3 See Pay Board, *Special Report; Relative Pay of Mineworkers*, Cmnd. 5567, 1974.

4 Before 1961, a surface shift consisted of $8\frac{1}{2}$ hours and an underground shift of $7\frac{1}{2}$ hours. In 1961 shift times at the surface and underground were reduced by $\frac{1}{4}$ hour. In 1969 hours at the surface were reduced by a further $\frac{1}{4}$ hour to 8.

5 Turner, 1964.

6 *Ibid.*

7 Using the D of E April earnings enquiry, Barratt Brown calculated the following figures for miners' weekly and hourly earnings relativities:

	1949	1956	1960	1966
Weekly	129	126	113	106.5
Hourly	140	149	143	142

In calculating hourly earnings he made an '*assumption* of, five shifts of 8 hours to 1961 and $4\frac{3}{4}$ shifts of $7\frac{3}{4}$ hours thereafter . . . the $4\frac{3}{4}$ shifts are the average number worked by those men presenting themselves for work in that week' (although he subsequently confirmed that he used $4\frac{3}{4}$ for the 1956 figure as well). These assumptions are inaccurate. The 8 and $7\frac{3}{4}$ hours per shift he used are simply the mean of surface and underground hours in a shift. This overestimates hours because it takes no account of the fact that a greater proportion of shifts are worked underground than at the surface. On the other hand, it takes no account of overtime hours, although the earnings figures include overtime pay. In every year except 1949, miners worked more hours than Barratt Brown assumed, and variations in hours between the years were also greater than between his assumed hours worked. There is a discrepancy between Barratt Brown's 1966 figure and our figure in Table 8.1 because, by using April, his figures do not catch the relatively depressing effect of the NPLA on earnings, whilst the October figures do. Even so, there is no evidence to suggest that miners maintained their relativity in hourly terms after 1960.

8 Net real income is calculated as annual gross money income adjusted for price changes and deductions for income tax and national insurance contributions. This procedure follows that of Frank Wilkinson and H. A. Turner in 'The Wage Tax Spiral and Labour Militancy' in Jackson, Turner and Wilkinson, *Do Trade Unions Cause Inflation?* University of Cambridge, Department of Applied Economics, Occasional Paper 36, CUP, 1972.

9 Standard errors in brackets.

10 If one replaces S_t with ΔS_t the coefficient is insignificant.

11 D of E, *Report of a Court of Enquiry into a Dispute Between the National Coal Board and the National Union of Mineworkers*, Cmnd. 4903, 1972.

12 *Ibid.*, paragraph 41.

13 Cmnd. 5567, 1974.

14 *Ibid.*

15 The wage relativities in this section are based on the Department of Employment, October Earnings Enquiry, and the monthly index of earnings, published in the *Department of Employment Gazette.*

16 For further details of the settlement, see Incomes Data Services, *Reports* 181–182, March and April 1975.

NOTES TO CHAPTER 9

1 NCB, *Annual Report*, 1966–67, pp. 36–7.

2 *Ibid.*, p. 36.

3 NCB, *Memorandum of Agreements*, Part 2, 20 March 1940, to 31 July 1946, pp. 129–31.

4 *Ibid.*

5 NUM, *Annual Conference*, Report and Proceedings, p. 9.

6 National Board for Prices and Incomes, *Coal Prices*, Report No. 12, Cmnd. 2919 February 1966.

7 NCB, *Annual Report*, 1966–67, para. 140.

8 NCB, Information Paper, No. 74, 'Productivity Trends'.

9 NCB, Financial Returns, 1963–73.

10 NUM, *Annual Conference*, Report and Proceedings, 1970.

11 NUM, *Annual Conference Report*, 1970, pp. 164–7.

12 Department of Employment, *Report of a Court of Inquiry into a Dispute Between the National Coal Board and the National Union of Mineworkers*, under the Chairmanship of the Rt. Hon. Lord Wilberforce, HMSO, Feb. 1972, Cmnd. 4903.

13 Department of Energy, Coal Industries Examination, *Final Report*, 1974.

NOTES TO CHAPTER 10

1 See H. A. Clegg, *The System of Industrial Relations in Great Britain*, second edition, Oxford, 1972, pp. 325–7; W. E. J. McCarthy, 'The Nature of Britain's Strike

Problem', *British Journal of Industrial Relations*, Vol. VIII, July 1970, pp. 224–38; Michael Silver, Recent British Strike Trends: A Factual Analysis', *British Journal of Industrial Relations*, Vol. XI, March 1973, pp. 66–104; and Richard Pryke, *Public Enterprise in Practice*, MacGibbon and Kee, London, 1971, pp. 95–6.

2 For an analysis of variations in strike frequency during this period, see K. G. J. C. Knowles, *Strikes*, Oxford, 1952.

3 G. B. Baldwin, *Beyond Nationalisation*, Chapter 3, pp. 86–7, Harvard University Press, 1955.

4 *Ibid.*

5 W. H. Scott, *et al.*, *Coal and Conflict*, Liverpool University Press, 1963.

6 Baldwin, *op. cit.*

7 Scott, *op. cit.*, p. 124.

8 See Chapters 3 and 4.

9 H. A. Clegg, *The System of Industrial Relations in Great Britain*, pp. 326–7, Second Edition, Oxford, 1972.

10 In practice, however, the completion-of-task bonus did tend to become a non-variable pay element.

11 The relevant dates ranged from 1951 in Yorkshire to 1965 in Kent.

12 There was a strike of winding enginemen in Yorkshire in 1956, for example, which included 84 coalmines and so inflated the dispute figure for Yorkshire in that year. There were also widespread restrictions of work in South Wales in 1955 and 1956, without which the incidence of disputes in those two years would have been lower and approximately equal to the frequency in 1957.

13 Our hypothesis that the average coalmining strike rate has responded at least in part to market factors, as reflected in the level of inland coal consumption over the post-war period, is tested in the following linear equation, which refers to the period 1950–73, differentiating between short-run and long-run variations in consumption.

$$WS_t = -6.052 + 0.038\ C_t + 0.056t \quad r^2 = 0.64$$
$$(2.159)\ (0.009)\ \ (0.037)$$

where WS = coal strikes per 1,000 miners (Department of Employment).
C = inland coal consumption in million tons, including the correction shown in Figure 10.4 for 1972 (*Digest of Energy Statistics*).
t = a time trend 1950–73. The bracketed figures are standard errors.

14 National Coal Board Dispute Returns 1965.

15 Dudley Jackson, H. A. Turner and Frank Wilkinson, *Do Trade Unions Cause Inflation?*, Cambridge University Press, 1972.

16 For an elaboration of this argument, see L. J. Handy 'The British Coal Miners' Strike in Perspective', *Economic and Political Weekly*, 4 March 1972, Department of Applied Economics Reprint No. 400.

17 For an analysis of the circumstances surrounding this claim, see L. J. Handy and K. J. Wigley, 'The Mines: Supply and Demand', *New Society*, 12 November 1970. Department of Applied Economics Reprint No. 344.

18 L. J. Handy and K. J. Wigley, *op. cit.*

NOTES TO CHAPTER 11

1 Pay Board, *Special Report: Relative Pay of Mineworkers*, Cmnd. 5567, HMSO, March 1974.

2 W. B. Reddaway, 'Wage Flexibility and Manpower Distribution', *Lloyds Bank Review*, No. 54, October 1959.

3 Council on Productivity, Prices and Incomes, *First Report*, para. 146, London, HMSO.

4 OECD, *Wages and Labour Mobility*, Paris, 1965.

5 *Ibid.*

6 Reddaway, *op. cit.*

7 OECD, *op. cit.*

8 See M. R. Fisher, *The Economic Analysis of Labour*, Ch. 8, pp. 214–25, for an elaboration of this point.

9 Both retirement and dismissal may be affected by wage and general labour market considerations, but it is doubtful if these are the primary factors influencing retirement and dismissal rates.

10 A worker may also be influenced by the lifetime earnings expectations offered by a particular employer or occupation. He is also likely to be particularly influenced in the short run by the training costs which he may have to bear, in the form of forgone income, until he attains the requisite skill for a particular job. In coalmining, for example, a new recruit would have to undergo eighteen months training before he would qualify for facework and, if he got it, the highest mining wage rate.

11 This obviously has certain inadequacies as an indicator of demand, as it does not reveal the very real differences in the supply and demand for particular skills at any moment. The general shortage of skilled workers, for example, makes it likely that craftsmen will be able to switch jobs fairly easily, despite a high general level of unemployment. This may also make them more sensitive to wage relativities.

12 The data were obtained from the Department of Employment, which provided the full flow matrices underlying the article which appeared in the *Gazette*, April 1970, pp. 303–7.

13 An alternative approach would have been to combine *RM* and *RC* into a single composite index of wage relativities; in view of the collinearity between the two this would hardly affect the result.

14 The corresponding equation using the mining/construction wage relativity was

$$\underset{}{VW} = 13.03 - \underset{(1.5)}{2.11}U_t - \underset{(0.7)}{0.14}RC \quad r^2 = 0.64 \quad DW = 1.28.$$

(1.5) (0.7) (0.02)

15 A figure supplied by the NCB.

16 MacKay, D. I. *et al.*, *Labour Markets under Different Employment Conditions*. University of Glasgow Social and Economic Studies, George Allen and Unwin, 1973.

17 Pay Board, *op. cit.*

18 *Digest of Energy Statistics* 1974.

19 NCB *Annual Report and Accounts* 1973–74 HMSO 1974.

20 *Digest of Energy Statistics*, 1974.

21 Data supplied by the NCB.

22 D of E, *Report of a Court of Enquiry into a Dispute Between the National Coal Board and the National Union of Mineworkers*, Cmnd. 4903, 1972.

23 Based on conversations with NCB manpower officers.

24 Following the analysis described in the text, the NCB manpower department has discovered that voluntary wastage of craftsmen is more sensitive to external wage relativities than it is for miners as a whole.

NOTES TO CHAPTER 12

1 Royal Commission on Trade Unions and Employers' Associations 1965–68, *Report*, June 1968, London, HMSO, Cmnd. 3623.

NOTES TO CHAPTER 13

1 These figures exclude allowances in kind as well as sickness and holiday entitlements.

2 Preamble to Incentive Scheme, November 1977.

3 *Ibid.* Clause (1).

4 In this context, 'standard task' means the agreed quantum of work specified in an installation agreement and includes any work directly related to advancing the coalface or drivage. There was some disagreement between the NUM and NCB on how task performance should be measured: the NCB wanted measurement in terms of square yardage while the NUM preferred measurement to be in lineal yards. The parties eventually agreed a compromise to the effect that performance would be measured in lineal yards at a standard depth of cut.

5 The *Guardian* 20 October 1978.

6 The *Yorkshire Post* 18 October 1978.

7 The *Yorkshire Post* 26 October 1978.

8 In fact, trial agreements on drivages had been operating in one of the NCB regions since 1974!

9 Reported in the *Daily Telegraph*, 9 November 1977. The NCB subsequently confirmed that they had been running a trial incentive scheme for 10 weeks at *Bolsover colliery* with the consent of the Nottinghamshire NUM. *Daily Telegraph*, 10 November 1977.

10 *Ibid.*

11 Figures supplied by the NCB.

12 From 1975 the grade structure was simplified. This did not affect the rates paid for any job but underground grades for face, development and other daywagemen were designated U1-U7. Faceworkers previously referred to as NPLA men and development workers previously referred to as Third Daywage Structure grade A men are now referred to jointly as U1 men. At the same time the grades for surface workers were redesignated S1-S6.

13 NCB statistics.

14 See Chapter 9.

15 NCB statistics.

BIBLIOGRAPHY

(A) STATISTICAL SOURCES

This study has been based largely on unpublished data. The published sources referred to were:

National Coal Board, *Report and Accounts*, 1946–76
Ministry of Fuel and Power, *Statistical Digest*, 1938–55
Ministry of Power, *Statistical Digest*, 1956–67
Ministry of Power, *Digest of Energy Statistics*, 1968–69
Department of Trade and Industry, *Digest of Energy Statistics*, 1970–73
Department of Energy, *Digest of UK Energy Statistics*, 1974–
Department of Energy, *Energy Trends*
Ministry of Power, *Statistical Returns on the Proceeds, Costs of Production, and Profits and Losses of Collieries*, 1938
Central Statistical Office, *Monthly Digest of Statistics*
Central Statistical Office, *Annual Abstract of Statistics*
Department of Employment, *Gazette*
Department of Employment and Productivity, *British Labour Statistics, Historical Abstract*, 1886–1968, HMSO, London 1971.
Department of Employment, *British Labour Statistics, Year Book*, 1969.
Department of Employment, *New Earnings Survey*, 1968
Department of Employment, *Time Rates of Wages and Hours of Work* 1948–
Incomes Data Services, *Reports and Bulletins*, 1967.

(B) WORKS CITED IN TEXT

Alexander, K. J. W. 'Wages in the Coalmining Industry since Nationalisation', *Oxford Economic Papers*, Vol. 8, No. 2, June 1956.
Arnot, R. Page. *The Miners*, Allen & Unwin, London 1949.
The Miners: Years of Struggle. Allen & Unwin, London 1953.
The Miners in Crises and War. Allen & Unwin, London 1961.
Baldwin, G. B. *Beyond Nationalisation: The Labour Problems of British Coal*. Harvard University Press 1955.
Barratt Brown, M. 'Determinants of the Structure and Level of Wages in the Coalmining Industry since 1956', *Bulletin of the Oxford Institute of Statistics*, Vol. 29, May 1967.
Beacham, A. 'The Coal Industry' in Burn D., *The Structure of British Industry*, Vol. 1, pp. 108–55 NIESR, CUP, London 1958.
Clegg, H. A. *The System of Industrial Relations in Great Britain*, second edition, Blackwell, Oxford 1972.
Cmd. 6364, HMSO, 1942.
Coal Industry Commission, *Interim Report*, Cmd. 359, HMSO, London 1919, *Final Report*, Cmd. 361, HMSO, London 1919 (Chairman Sankey).
Report (Chairman Samuel), Cmd. 2600, HMSO, London 1925.

Council on Productivity, Prices and Incomes, *First Report*, HMSO, London 1959.

Department of Employment, *Gazette*, April 1970, HMSO.

Report of a Court of Enquiry into a Dispute Between the National Coal Board and the National Union of Mineworkers (Chairman the Rt. Hon. Lord Wilberforce), Cmnd. 4903, HMSO, London 1972.

Department of Energy, Coal Industries Examination, *Final Report*, 1974.

Financial & Economic Obligations of Nationalised Industries, Cmnd. 1337, HMSO, London 1961.

Fisher, M. R. *The Economic Analysis of Labour*, Weidenfeld and Nicolson, London 1971.

Handy, L. J. 'Absenteeism and Attendance in the British Coalmining Industry since Nationalisation', *British Journal of Industrial Relations*, Vol. VI, No. 1, 1968.

'The British Coal Miners' Strike in Perspective', *Economic and Political Weekly*, 4 March 1972.

Handy, L. J. & Coutts, K. J. 'Pay and Pit Wastage', *New Society*, January 1974.

Handy, L. J. & Wigley, K. J. 'The Mines, Supply and Demand', *New Society*, 12 November 1970.

Jackson, D, Turner, H. A. & Wilkinson, F. *Do Trade Unions Cause Inflation?* CUP, 1972.

Knowles, K. G. J. C. *Strikes* Blackwell, Oxford 1952.

McCarthy, W. E. J. 'The Nature of Britain's Strike Problem', *British Journal of Industrial Relations*, Vol. VIII, July 1970, pp. 224–38.

MacKay, D. I. *et al. Labour Markets Under Different Employment Conditions*, University of Glasgow Social and Economic Studies, George Allen & Unwin, 1973.

Ministry of Fuel and Power, *Report of Technical Advisory Committee on Coalmining* (Chairman Reid), Cmd. 6610, HMSO, London 1945.

Coal Industry Nationalisation, Cmd. 6716, HMSO, London 1945.

Ministry of Power, *The Finances of the Coal Industry*, Cmnd. 2805, HMSO, London 1965.

Fuel Policy, Cmnd. 2798, HMSO, London 1965.

Fuel Policy, Cmnd. 3438, HMSO, London 1967.

National Board for Prices and Incomes, *Coal Prices*, Report No. 12, Cmnd. 2919, HMSO, London 1966.

Payment by Results Systems, Report No. 65, Cmnd. 3627, HMSO, 1968 and Supplement, Cmnd. 3627-1, HMSO, 1968.

Coal Prices, Report No. 124, Cmnd. 4149, HMSO, London 1969.

Coal Prices, 2nd Report No. 153, Cmnd. 4455, HMSO, London 1970; Supplement No. 2, Cmnd. 4455-2, HMSO, London 1971.

National Coal Board, *Report of the Advisory Committee on Organisation*, (Chairman Fleck), NCB, London 1955.

Memorandum of Agreements, Arbitration Awards and Decisions, Recommendations and Interpretations Relating to National Questions Concerning Wages and Conditions of Employment in the Coalmining Industry of Great Britain, Parts 1-30, NCB, London.

Annual Report and Statement of Accounts, NCB, London 1946–75.

Productivity Trends, NCB Information Paper No. 74, London 1968.

Coal News, NCB Headquarters, London 1966–75.

Investing in Coal, NCB London 1956.

Plan for Coal, NCB London 1950.

Revised Plan for Coal, NCB London 1959.

National Union of Mineworkers, *Rule Book*, NUM, London 1972.

Annual Conference, Reports and Proceedings, 1946–75, NUM, London.

Minutes of National Executive Committee and Reports of Special Conferences, 1946–75, NUM, London.

Miner, NUM, London 1969–

OECD, *Wages and Labour Mobility*, Paris 1965.

Pay Board, *Special Report: Relative Pay of Mineworkers*, Cmnd. 5567, HMSO, London 1974.

Penrice, G. A. 'Earnings and Wage Rates Since 1938', *London and Cambridge Economic Service Bulletin*, September 1952.

'Earnings and Wage Rates 1948–55', *London and Cambridge Economic Service Bulletin*, December 1965.

Pryke, Richard, *Public Enterprise in Practice*, MacGibbon and Kee, London 1971.

Reddaway, W. B. 'Wage Flexibility and Manpower Distribution', *Lloyds Bank Review*, No. 54, October 1959.

Rowe, J. W. F. *Wages in the Coal Industry*, King, London 1923.

Royal Commission on Trade Unions and Employers' Associations, *Minutes of Evidence* 4 (NCB), HMSO, London 1966.

Royal Commission on Trade Unions and Employers' Associations, 1965–68, *Report*, (Chairman Donovan), Cmnd. 3623, HMSO, London 1968.

Sales, W. H. & Davies J. L. 'Introducing a New Wage Structure into Coalmining', *Bulletin of the Oxford University Institute of Statistics*, Vol. 19, No. 3, 1957.

Scott, W. H. *et al. Coal and Conflict*, Liverpool University Press, Liverpool 1963.

Searle-Barnes, R. G. *Pay and Productivity Bargaining*, Manchester University Press, Manchester 1969.

Silver, Michael. 'Recent British Strike Trends: A Factual Analysis', *British Journal of Industrial Relations*, Vol. XI, March 1973, pp. 66–104.

Turner, H. A. 'Trade Union Differentials and the Levelling of Wages', *Manchester School*, September 1952.

'Wages: Industry Rates, Workplace Rates and the Wage Drift', *Manchester School*, May 1956.

'More on the Wages Drift', *Manchester School*, January 1960.

'The Disappearing Drift (Or in Defence of Turner)', *Manchester School*, May 1964.

Wilson, H. *New Deal for Coal*, Contact, London 1945.

(C) OTHER RELEVANT WORKS

Brown, William A. 'Piecework Bargaining', Warwick Studies in Industrial Relations, 1973.

Cole, G. D. H. *Labour in the Coalmining Industry, 1914–21*, Clarendon Press, Oxford 1923.

Court, W. H. *Coal*, Longmans, Green & Co., London 1951.

Dennis, M., Henrigues, F. and Slaughter, C. *Coal is Our Life*, Eyre and Spottiswoode, London 1956.

Dickie, J. P. *The Coal Problem, A Survey: 1910–1936*, Methuen, London 1936.

Haynes, W. W. *Nationalisation in Practice: The British Coal Industry*, Bailey & Swinfen, London 1953.

Heinemann, M. *Britain's Coal: A Study of the Mining Crisis*, Victor Gollancz, London 1944.

Horner, A. L. *Incorrigible Rebel*, MacGibbon & Kee, 1960.

Hughes, J. & Moore, R. (eds.) *A Special Case?*, Penguin, Harmondsworth 1972.

Jones, Hugh E. M. *Wage Structure in Theory and Practice*, North Holland, 1966.

Kelly, D. M. & Forsyth, D. J. C. (eds.) *Studies in the British Coal Industry*, Pergamon, Oxford 1969.

Moffat, A. *My Life with the Miners*, Lawrence and Wishart, London 1965.
Paynter, W. *British Trade Unions and the Problems of Change*, Allen & Unwin, 1970.
My Generation, Allen & Unwin, 1972.
Platt, J. *British Coal: A Review of the Industry, Its Organisation and Management*, Lyon, Grant and Green, London 1968.
Raynes, J. R. *Coal and Its Conflicts: A Brief Record of the Disputes Between Capital and Labour in the Coalmining Industry of Great Britain*, Ernest Benn, London 1928.
Redmayne, R. A. S. *The British Coalmining Industry During the War*, Clarendon Press, Oxford 1923.
Reid, G. L., Allen, K. & Harris, D. J. *The Nationalised Fuel Industries*, Heinemann, London 1973.
Reynolds, Lloyd G. & Taft, Cynthia H. *The Evolution of Wage Structure*, New Haven 1956.
Robinson, C. *Competition for Fuel*, Institute of Economic Affairs, London 1971.
Robinson, D. (ed.) *Local Labour Markets and Wage Structures*, Gower Press, London 1970.
Select Committee on Nationalised Industries, *Nationalised Coal Board,* HMSO, London 1969.
Simpson, E. S. *Coal Power Industries in Post-War Britain*, Longmans, London 1966.
Townsend, Rose H. *The British Coal Industry*, Allen & Unwin, London 1951.
Zweig, F. *Men in the Pits*, Victor Gollancz, London 1949.

INDEX